ENERGY POVERTY IN EASTERN EUROPE

To Dimitrije and Eleni, my parents

Energy Poverty in Eastern Europe
Hidden Geographies of Deprivation

STEFAN BUZAR

University of Birmingham, UK
Visiting Professor of Economic Geography, University of Gdansk, Poland

Routledge
Taylor & Francis Group

LONDON AND NEW YORK

First published 2007 by Ashgate Publishing

Published 2016 by Routledge
2 Park Square, Milton Park, Abingdon, Oxfordshire OX14 4RN
711 Third Avenue, New York, NY 10017, USA

First issued in paperback 2016

Routledge is an imprint of the Taylor & Francis Group, an informa business

British Library Cataloguing in Publication Data
Buzan, Stefan
    Energy poverty in Eastern Europe : hidden geographies of
    deprivation
    1. Energy consumption - Europe, Eastern 2. Poverty -
    Europe, Eastern
    I. Title
    333.7'96313'0947

Library of Congress Cataloging-in-Publication Data
Buzar, Stefan, 1977-
    Energy poverty in Eastern Europe : hidden geographies of deprivation / by Stefan Buzar
        p. cm.
    Includes bibliographic references and index.
    ISBN-13: 978-0-7546-7130-5
    1. Energy policy--Europe, Eastern. 2. Power resources--Europe, Eastern. 3. Energy
policy--Europe, Eastern--Case studies. 4. Power resources--Europe, Eastern--Case
studies. 5. Europe, Eastern--Economic integration. I. Title.

    HD 9502.E83B89 2007
    333.790947--dc22

                                                                    2007011103

ISBN 13: 978-1-138-27635-2 (pbk)
ISBN 13: 978-0-7546-7130-5 (hbk)

Contents

List of Figures

List of Maps and Plates

List of Tables

Foreword

When I first started working on energy poverty in Eastern Europe six and a half years ago, I could hardly imagine the immensity of the iceberg that I had stumbled across. With the naivety of a young researcher, I simply decided to pursue a line of work which, it seemed at the time, could help me make an easy transition from green political activist to social scientist, while allowing me to concentrate on a subject that I strongly cared about. A subject that was then virtually unknown in continental Europe, let alone Eastern Europe and the former Soviet Union. This group of countries, I felt, urgently needed an academic study of energy poverty, not only because of the cold climates, rising poverty rates and dilapidated housing, but also because their governments were being made to believe that they could solve the energy-related problems inherited from the Communist era solely by raising energy prices and liberalising their energy sectors along the Thatcherite model implemented in Britain. Broader social, institutional and spatial issues were simply not on the agenda, although it was precisely the British experience that had shown how the neglect of such problems can lead to more severe problems at a later stage.

I soon discovered, however, that studying energy poverty in Eastern Europe would be anything but simple. The first obstacle was the lack of data: the policy makers' lack of awareness of the issue had translated into a paucity of statistical surveys and standardized data. But that was just the beginning. Once I found indirect ways of getting the information I needed, I began to realize that energy poverty was far more complex than I initially thought, that it had multiple facets which extended like tentacles throughout the economy, society and the built environment. It soon became apparent that many broader and better publicized problems faced by the countries I studied – general non-payment for energy services, the financial insolvency of energy companies, inadequate investment in building maintenance, social exclusion and marginalization – were all directly linked to this problem. The culmination, however, must have happened when I realized that energy poverty in Eastern Europe was, and still is, affecting the citizens of Western Europe and other neighbouring regions in many unexpected ways. For example, every time Russia sends shockwaves of panic throughout the EU by temporarily shutting off the oil and/or gas supply to one of its neighbouring countries, the social and political implications of energy price increases are somehow to blame. As argued by the book, inadequately heated homes are one of the main cogs in the machine of domestic energy deprivation, social policy, housing reform and energy prices in former Soviet Republics and other post-socialist states.

Given its multiple implications, then, it is striking how little attention energy poverty has received not only in the Western polity, but also the Eastern European countries that are directly affected by it. The concept of 'energy poverty' is still absent from the mainstream language of decision-makers and experts in Europe. It is virtually unknown to the relevant academic and policy literatures.

Still, the picture is not entirely bleak. In recent years, the expert community has started to take upon the multiple theoretical and conceptual challenges posed by the problem: there have been conferences, policy reports, and academic articles. Some policy steps have been undertaken, too: for example, the Energy Community treaty, signed by eight Balkan countries under the oversight of the EU, was the first international document of its kind to explicitly incorporate a Memorandum on Social Issues.

A lot remains to be done in the coming years. We lack an integrated theoretical understanding of the problem – the relationships between energy efficiency, institutional cultures, social policy and cold homes require further elucidation. Moreover, there are no standardized measurement frameworks for energy poverty and no consistent systems of data gathering, which makes it difficult for experts to undertake cross-sectional studies, and activists to increase public awareness. Most importantly, perhaps, there is insufficient knowledge about the kinds of policy frameworks needed to address the problem not only in the Central and Eastern European countries where it engulfs large parts of the population, but also in the post Soviet states further East which bask in the safety of cheap oil and gas without noticing the ominous shadow of domestic energy deprivation on the horizon. I hope that this book can provide at least the first step towards addressing some of these issues.

Stefan Buzar (Bouzarovski),
Oxford, 2007

Acknowledgements

This book grew out of a doctoral thesis that was financially supported by a Scatcherd Scholarship from the University of Oxford, a Larkinson Scholarship from St. Hugh's College, Oxford, and an Overseas Research Award. I don't know how I could have survived the trials and tribulations that usually accompany a doctoral degree without the kind coaching and encouraging criticisms of my two supervisors: Dr Judy Pallot and Dr Brenda Boardman. I am also indebted to the examiners – Professor Gordon Clark and Dr Alison Stenning – for their insightful comments to the final draft of the thesis. Since October 2004, a Junior Research Fellowship at Christ Church, Oxford, has given me the necessary time and space to extend the results of the thesis into academic papers and further research that produced this book.

Much of the background work for the book was undertaken in the Czech Republic and Macedonia, and it is in this instance that I would like to acknowledge the kind assistance of Jan Dušik (Czech Ministry of Environment), as well as the World Carfree network (France, Ivana, Juho, Petr, Randall, Richard), who provided me with the infrastructure for field research. In addition, I am grateful to Martin Dašek, Eva Holánová, Jarmila Škvrňová, Martin Kloz, Dr Martin Lux, Jaroslav Kubečka and Milada Podpěrová for their help with the interviews. In Macedonia, the data collection process was undertaken with the support of Professor Stefanka Hadži Pecova, Dr Elisaveta Stikova, Trajče Čerepnalkovski, Lile Atanasovska, Ivica Sekovanik', Milka Petkovska, Penka Nikolovska, Pene Penev, Slobodan Apostolov, Vesna Trajkova and Andrijana Veljanoska, as well as the Barbinder-Watson Trust Fund at St. Hugh's College. I will never forget the kind-hearted support of Jelica Trpčevska, who sadly is no longer with us.

Besides helping me with the data gathering, Saška was one of the few people who was always there, always ready to bear the brunt of bad moods or help me get through difficult moments. During the past 5–6 years, friends such as Aneta, Biljana, Daniela, Elena, Elton, Francesca, Maria, Ivan, Iwona, Jeanette, Kathryn, Maja, Martin, Mojca, Murat, Olga, Oliver, Richard, Roza, Sandra, Stach and others, have made my life brighter, giving me the motivation to complete this project. But in the end, the whole undertaking would have been impossible without the unquestioning support of my loving parents (Dimitrije and Eleni), sister (Rumena), and grandmother (Marika).

Chapter 1

Setting the Framework

While the challenge in most of the developing world is to expand infrastructure and service delivery, East European and Central Asian countries are struggling to prevent the existing energy systems from failing (World Bank, 2005)

Ukraine is at the verge of a second energy crisis (REGNUM News Agency, 14 February 2006)

The ex-Communist states of Eastern and Central Europe (ECE) and the former Soviet Union (FSU) encompass a population of more than 400 million people and an area that covers nearly 20 per cent of the world's landmass. Since the early 1990s, they have been undergoing a process of deep systemic change, aimed at reforming the economic, political and social structures inherited from the Communist era. The new geopolitical situation has forced these countries 'onto the economic geography agenda, not the least because the map of this part of the world has now been redrawn and there are many new states to be integrated into the world economy' (Stenning and Bradshaw, 1999, p. 97).

However, the sheer extent of the post-socialist reform process has meant that social scientists and policy-makers have often neglected some of its finer aspects. The consequent accumulation of overlooked problems across inter-connected sectors of the economy has often led to severe social and political problems. One such situation exists at the boundary of the energy, housing and social welfare domains, where the slow pace of reforms has created major social and infrastructural difficulties. Winter after winter, the world's media are replete with reports about energy-related humanitarian emergencies in post-Communist Eastern Europe and Asia. 'Georgia's president has cut short a visit to the World Economic Forum in Switzerland and headed home to deal with a growing energy crisis' stated the BBC news website on the 29th of January 2006, while just a week earlier it was reported that 'schoolchildren are missing classes and boiler faults have left some households shivering as Arctic cold grips Russia' (17 January 2006). It seems that every spell of unusually cold weather brings misery to countries in the region, as millions are left without heating and electricity, due to the energy utilities' financial and technical problems (Lovei et al., 2000; Timofeev, 1998).

At the same time, post-socialist reform experts have been talking about a second type of energy crisis. Many countries in the region, especially those in ECE, have recently undertaken significant energy price increases, with the aim of removing the economic structure inherited from socialism. Before 1990, tariffs were set at below cost-recovery levels and there were extensive cross-subsidies from industry to the residential sector. The problem that has emerged in the post-socialist transition, however, is that most governments have been unable to develop the necessary social safety net to protect vulnerable households from energy price increases. This leaves many families with no option other than to cut back on their energy purchases. For example, a cross-country report in the Balkans has argued that power affordability is a problem 'for many consumer groups' in this region, including 'pensioners, unemployed, low income households'. Moreover 'many of the South East European countries have not yet developed adequate social safety mechanisms to protect energy poor consumers' (EBRD, 2003).

The key premise of this book is that the two post-socialist energy crises are underpinned by a common predicament – the 'hidden geography' of energy poverty. By definition, a household is energy poor if it is living in an inadequately heated home, which can mean that either the average daytime indoor temperature of the dwelling is below the biologically-determined limit of 20°C necessary to maintain comfort and health (Boardman, 1991), or that the amount of warmth in the home is lower than the subjective minimum which allows an individual to perform his/ her everyday life. There is a wide body of evidence to suggest that ECE and FSU countries are facing an escalating energy poverty problem, due to their specific social and physical conditions, such as: cold climates, temperature inversions, the removal of universal socialist-era energy price subsidies and falling real incomes post-1990 (Lampietti and Meyer, 2003). Although it has been shown that energy poverty in other countries, such as the UK, emerges through an interaction of low incomes, institutional strategies and housing conditions (Rudge and Fergus, 1999; Wicks and Hutton, 1986; Bradshaw and Harris 1983, Burghes, 1980), this theoretical connection has yet to be made in the post-socialist context.

Why energy poverty matters

> Eradicating poverty is the greatest global challenge facing the world today and an indispensable requirement for sustainable development, particularly for developing countries ... This would include actions at all levels to: ... the access of the poor to reliable, affordable, economically viable, socially acceptable and environmentally sound energy services (World Summit for Sustainable Development, Johannesburg, 4 September 2002).

'Poverty reduction' and 'improved energy services for the poor' have become the buzzwords of global development discourses during the past two decades. The world's key multilateral development institutions, including the United Nations' Development Programme, the World Bank, UK Department for International Development – as well as influential international NGOs such as the Intermediate Technology Development Group, Friends of the Earth and Greenpeace – have placed a heavy emphasis on 'sustainable energy for poverty reduction'. A wide range of financing instruments has been developed to support this goal, mostly by facilitating 'technology transfer' into developing countries (Buckley et al., 2001; World Bank, 1993).

However, although it has become commonplace to refer to 'energy poverty' in the context of developing countries in the South, the concept has yet to be related to the emergent socio-economic realities of the post-socialist states of the global North. Not a single academic or policy paper published to date has dealt explicitly with the problem of energy poverty in Eastern Europe. This is despite the growing number of studies of the social implications of energy reforms in post-socialist countries (see, for example Kovačević, 2004; Velody et al., 2003; Lampietti and Meyer, 2003; Dodonov et al., 2001; Lovei et al., 2000), which have, through different analytical instruments, provided a wide range of views about the relationships between energy tariffs, affordability and post-Communist reforms. In the European context, the only academic contribution that provides an integrated perspective on the issue is Healy's (2004) book on *Housing, Fuel Poverty and Health: a Pan-European Analysis*, which contains a cross-country overview of the problem. However, Healy sidesteps the Eastern European countries where energy poverty is a much bigger issue. It can be argued that the failure to conceptualize energy poverty as a distinct phenomenon, with specific contingencies and implications, has led to the theoretical marginalization of the housing, social policy and governance dimensions of energy reforms.

As will be discussed later in this book, the attitude of many policy-makers and experts in the ECE region is reminiscent of the situation in the United Kingdom during the early 1970s, when the notion of a distinct energy poverty problem was intercepted by a government minister with the statement: 'People do not talk of 'clothes poverty' or 'food poverty' and I do not think that it is useful to talk of 'energy poverty' either' (Boardman, 1991, p. 1). His view is now consigned to history, as it has been overruled by the immense body of evidence about the health problems faced by tens of thousands of British households living in poorly heated homes (*ibid.*, and Wynn and Wynn, 1979; Bradshaw and Harris, 1983; Lewis, 1982; Isherwood and Hancock, 1979). Today, 'fuel poverty' is part of the mainstream of policy and academic discourses in the UK.

But will this be the case in the FSU and CEE? At the July 2006 G8 summit in St Petersburg, Russian Finance Minister Sergey Kudrin 'noted the importance of fighting against energy poverty', because 'it is impossible to develop the economy, improve health care and develop education without having access to energy resources' (RIA Novosti, 2006). Such statements, which were also embedded in the final declaration of the summit, echo President Putin's earlier announcement that 'Russia intends to come up with concrete initiatives and proposals on how to confront energy poverty, improving diversification and ensuring energy conservation security during the forthcoming G8 summit' (Russia & CIS Business & Financial Daily, 2006). The judgement is still out as to whether and how this objective will be applied to the post-socialist context itself.

An integrated view of energy poverty in the ECE and FSU

One of the key underlying problems in the conceptualization of energy poverty in post-socialism is the lack of an integrated theoretical understanding of the interdependencies of social, energy and housing reforms. The restructuring process has turned the regions east of the Elbe and the Alps into a 'playground of history' (Przeworski, 1991, p. 21), where socio-economic processes unfold and differentiate at unprecedented speeds. These changes have brought about new trends in 'the geographies of the region in the forms of metropolitan growth, the economic collapse of peripheral regions, the polarization of urban spaces as inequalities deepen, and the reworking of the territorial structure and democratic spaces of the state and civil society' (Smith and Pickles, 1998, p. 5). In particular, 'unemployment soared ... income inequality has grown (in many countries it has exploded) and hardly any economic or social indicators have remained unaffected' (Fajth, 1999: 417). This is contrary to the initial optimism about post-socialist reforms, which expected that the economic benefits of the transformation process would quickly 'trickle down' to the majority of the population.

The transformation process has also created new geographies of unevenness at the transnational scale: 'while some countries have so far survived economic transition in reasonably good shape, for others the consequences of economic transition have been disastrous' (Redmond and Hutton, 1999, p. 1). Such divisions have deepened even further with EU enlargement, resulting in several regional groupings of countries with different levels of development. The 10 states that have joined the EU are, to various degrees, well integrated within the sphere of Western capitalism, while others in the Balkans and especially the FSU are still struggling with the problems of post-Communist transition. The widening gap between the two groups of countries has led some experts to declare that 'the similarities of systemic change have disappeared, and so has the research subject 'transition' (von Hirschhausen and Wälde, 2001: 107). However, 'economic transformation cannot be separated from changes in social, political and cultural realms, and must be examined within its regional, national and international context' (Stenning, 1997, p. 160). In this book, the post-socialist space is treated as a whole and the reform process is mainly described by the term 'transformation', because the more commonly used 'transition' implies a one-way movement towards a predefined state. The idea of 'transition' to a 'market economy' contradicts realities on the ground, which suggest a complex political and economic 'transformation' of the former socialist countries of ECE and FSU (Bradshaw and Stenning, 2000, p. 12; Pickvance, 1997; Smith, 1997).

The extensive social effects of the transition have been supplemented by the problematic situation in the energy sectors of nearly all ECE and FSU states. Despite being part of one of the most dynamic aspects of the economy, energy operations have presented a major reform challenge in the post-socialist period. Most former socialist countries pledged to adopt neoliberal energy legislation in the early 1990s (Wälde and von Hirschhausen 1998; Stern, 1994), requiring the vertical and horizontal unbundling of formerly state-owned integrated energy monopolies, the liberalization of energy markets and prices and the establishment of independent regulatory bodies. However, the high social and political costs of the energy reform process have delayed its implementation in the 'advanced' and the 'lagging' post-socialist states alike. Even leading reformers such as the Czech Republic have been reluctant to open up their energy markets in full or to privatize the national electricity monopoly (Kočenda and Cábelka, 1999). In some of the poorer post-Communist states, governments have been unable to compensate low income households with the adequate level of social protection. This has often resulted in bill recovery problems, as consumers have been unable or unwilling to pay for the electricity, gas or hot water provided by energy utilities (World Bank, 1999a).

There is also a deep disconnection between the literatures focussing on the spatial and institutional dimensions of post-socialist reforms, on the one hand, and the emergence of poverty in transformation, on the other. Although it is without doubt that the transformation process has exposed the post-socialist states to powerful globalizing tendencies, the nature of this relationship in the sphere of social exclusion remains unclear. For example, to what extent can we characterize post-socialist urban poverty as 'a set of spaces of juxtaposed fragments and contrasts, where diverse relational webs might coalesce, interconnect or disconnect' (Amin, 1999, p. 43)? Also, how is the post-socialist welfare state responding to the 'global crisis of Fordism' (Pierson, 1999, p. 423)? Unravelling the territorial and institutional background of domestic energy deprivation may help us find the answers to such questions. This means that the condition of energy poverty is thus significant in theoretical terms, because it encapsulates the broader interactions of economic, institutional and technical infrastructures.

At a more practical level, and perhaps most importantly of all, there is an urgent need for developing an analytical framework to interpret the growing body of evidence – both narrative and analytical – which suggests that millions of households in the region are living within a 'hidden' geography of domestic energy poverty. It is also necessary to determine how energy poverty relates to broader social exclusion patterns *per se*. This can provide local and national decision-makers with some of the badly needed expert knowledge for remedial action, if only by bringing the issue to the fore and stressing its innately path-dependent nature. Otherwise, the situation can only get worse if present trends of increasing residential prices, despite stagnant incomes, continue into the future. As evidenced by the experiences of other countries in Western Europe (see Healy, 2003), the economic and health costs of inaction can be prohibitively high.

Towards a geography of poverty

Adding to the analytical gaps in theoretical and policy understandings of energy poverty is the lack of a distinctively geographical conceptualization of poverty in the given context. A major part of post-socialist poverty research is produced and driven by welfare economists, and as such is disproportionately geared towards macroeconomic analyses. Nearly all the studies used by decision-makers in the region (the World Bank, IMF, EU, national governments) operate with a language of incomes, prices and poverty lines. More often than not, poverty is measured by dividing society into 'poor' and 'non-poor' individuals, or assuming a one-to-one relationship between income and welfare. The reliance on such blunt methodological tools is contradictory to the growing realization that 'discussions about 'advanced marginality' in network societies ... need to broaden their current scope of labour markets, techno-economic change, welfare restructuring ... and housing and economic restructuring' (Speak and Graham, 2000: 1991). A further problem is posed by the fact that very few studies extend beyond the domain of large scale economic and political interactions, into the lived experiences of poverty at the household level (for a further discussion see Cousins 2001, Burawoy and Verdery, 1999; Smith and Swain, 1998).

There is a growing awareness that poverty and social exclusion arise out of a combination of mutually-embedded and -networked conditions (Castells, 2002; Sibley, 2001; Blake, 2001; Graham, 2000). This implies that the dominant mode of poverty research will have to broaden and deepen its analytical horizons, if it is to foster more meaningful and effective practical solutions for social exclusion and marginalization. Such a project could involve the use of economic geography in 'foregrounding cultural and institutional variables previously thought out of bounds' to the development of social and economic structures (Clark et al., 2000, p. 5). Although geography has seen a resurgence of poverty research in the period since Leyshon (1995) noted the absence of this term from the third edition of *The Dictionary of Human Geography*, many of the finer institutional and spatial aspects of social exclusion remain insufficiently explored within the discipline (see Mohan, 2000; Byrne, 1999). There is a need for a greater theoretical and practical role of human geography within poverty studies.

Aims, arguments, methods, scope ...

In light of the above, the main purpose of this book is to contribute to an improved understanding of energy poverty in post-socialism. It aims to connect the literatures about the spatial and institutional dimensions of post-Communist economic reforms, on the one hand, and the emergence of poverty in transition, on the other. In this, it should foster an increasing awareness of the importance of organizational and social processes in constructing the spaces of social exclusion, while showing that 'the social is spatially constructed, too' (Massey, 1984, p. 6). I argue that energy poverty arises out of the inadequate co-ordination of energy, social welfare and housing policies. The key problem is the lack of properly targeted social assistance programmes, as well as the inadequate support frameworks for energy efficiency in the domestic sector.

The book also aims to emphasize the non-conformity of energy poverty with income based inequality measures, as the number of households suffering from energy expenditure problems may not match official poverty lines defined by the state. This is because the amount of useful warmth in the home is determined by a complex set of factors that are additional to income. These include the quality of the housing stock and heating systems, as well as the spatial and temporal distribution of daily occupancy patterns. Thus, aside from income, energy poverty amelioration policies must also address technology and housing stock issues. Such a line of reasoning expresses a desire to progress beyond reductionist interpretations of poverty *per se*. It accentuates the way in which deprivation is contingent on multi-scalar path-dependencies, state industrial policy and the interaction of fixed and mobile infrastructure networks.

Another aim of the book is to examine the institutional and spatial production of energy poverty. I am interested to know how household-level deprivation arises out of broader organizational and political relations. This investigation requires a greater level of analytical detail, and consequently I have chosen to focus on processes within Macedonia and the Czech Republic – two countries that have

Map 1.1   **Location of the study countries in the wider European context.**
Post-socialist states are shaded in grey

followed different development paths during the last 15 years – to illustrate broader trends (their locations and salient features are indicated in Maps 1.1 and 1.2). The analysis of the role of space in creating and shaping domestic energy deprivation predisposes the book towards a review of the territorial structure of energy poverty at different scales. I focus on the lived experiences of scarcity and marginalization at the household level, and the broader socio-demographic factors that distinguish energy poor families from the rest of the population.

These aims have been addressed with the aid of a variety of methods, including semi-structured interviews with policy-makers, professionals and households, as well as reviews of locally published literatures. The demographic size and structure of energy poverty has been examined with the aim of quantitative analyses of income and expenditure patterns (for further methodological information, see Fankhauser and Topic, 2005; Hulchanski, 1995), subjective perceptions of well-being (Townsend, 1979) and assessments of housing quality (Rudge and Nicol, 1999; Lewis, 1982). Further explanations of the particular set of methods and data sources in different parts of the analysis are available in the introductory parts of each chapter.

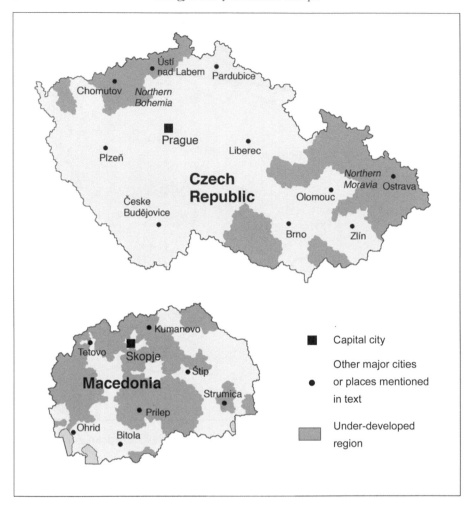

Map 1.2   Main cities in the two study countries, and their location within the broader ECE region

Source: CEA, 2005; MLSA, 2005.
Note: Areas with unemployment rates above the respective national averages (approximately 35 per cent in Macedonia, 10 per cent in the Czech Republic) are marked in grey.

In geographic terms, the book is focussed on the post-socialist states of Eastern and Central Europe (ECE) and the Former Soviet Union (FSU), although there is a heightened emphasis on the former. This is because ECE states are generally more 'advanced' in the post-socialist transformation process and are facing a wider range of issues as a result. Also, there is a greater availability of data for this region.

Defining the 'hidden' spaces of energy poverty

Coming up with a working definition of energy poverty was one of the most difficult tasks in the period leading up to the field research that produced this book. This is because there is hardly any consensus in the literature as to what constitutes poverty itself, let alone energy poverty (for a further discussion, see Boardman, 1991; Healy, 2003). In order to devise a framework that could be applied in the given context, I turned to the 'relative' definition of poverty, which is one of the most widely used approaches (see Percy-Smith, 2000; Townsend, 1979; Barr, 1998). According to this paradigm, poverty *per se* is seen as 'a lack of access to resources and denial of opportunities' which hampers an individual's ability to participate in the 'lifestyles, customs and activities which define membership of society' (Folwell, 1999, p. 5). Conceptualizing poverty in a 'relative' way opened the space for interpreting energy poverty through what Healy (2003, p. 36) terms the 'consensual' approach, which aims to capture the 'wider elements' of domestic energy deprivation, such as 'social exclusion and material deprivation, as opposed to approaches based solely on home-heating expenditure or household temperature'. This framework also benefits over other methods 'in that it is based on the households' actual feelings and statements … as opposed to being based solely on arbitrary calculations or estimations' (ibid.).

I thus defined energy poverty as *the inability to heat the home up to a socially- and materially-necessitated level*. A household is considered energy poor if the amount of warmth in its home does not allow for participating in the 'lifestyles, customs and activities which define membership of society'. Such a definition comprises both the biologically determined limit of 20°C necessary to maintain comfort and health (Boardman, 1991, but also see Rudge and Nicol, 1999; Healy, 2003) and the subjective minimum below which an individual feels unable to perform his/her everyday life. The advantage of this conceptualization, I believe, is that it takes into account the arguments of authors like Boardman, 1991 and Rudge and Nicol, 1999, who have claimed that prolonged exposure to temperatures below 20 or 21°C can be harmful over the long term, because breathing cold air affects the respiratory (and to some extent circulatory) system, regardless of the amount of clothing and humidity in the home. However, it also implies that a household may suffer from energy poverty even if its domestic temperature is above the biologically determined limit, provided that the temperature in the home is insufficient for performing usual social customs and practices. The presence of energy poverty can be detected through subjective surveys of well-being or patterns of household expenditure, as data about domestic temperature levels is almost nonexistent (Healy, 2003; Rudge and Nicol, 1999; Lewis, 1982; Townsend, 1979).

It should be noted that the book's definition of energy poverty is limited to the home and does not include mobility and residence outside the domestic domain (for a broader discussion of the definition of energy poverty, see Boardman, 1991; Healy, 2003). Energy poverty is thus closely connected to domestic energy efficiency, which is understood to comprise the ratio between the 'raw' energy arriving at the perimeter of the home, on the one hand, and the actual energy service received by the household inside it through lighting and heating appliances, on the other. The lesser the loss of energy during this conversion, the greater the energy efficiency of the home. Domestic energy efficiency thus reflects the technical quality of the building, and the ability of the built fabric, energy distribution installations and domestic appliances to minimize energy losses during the generation of useful warmth.

At this stage it is important to make a distinction between the concepts of 'energy efficiency' and 'energy conservation'. As pointed out by Boardman (1991) 'Invariably, to obtain energy efficiency improvements there has to be investment in capital stocks' (p. 3). She argues that 'it is virtually impossible to obtain improved energy efficiency through behavioural changes alone' since 'there would have to unused capital equipment already available' (*ibid.*). However, energy losses through the built fabric or heating systems can also be reduced through energy conservation measures, which do not require capital investment in the housing stock. For example, actions like turning off light in unused rooms or closing curtains and window shutters during cold weather would fall under the heading of energy conservation. Buying new and more efficient light bulbs, curtains or shutters, however, is an energy efficiency step.

The notion of a 'hidden geography' of poverty has provided a key conceptual apparatus for capturing the political and physical invisibility of post-socialist energy poverty in ECE. Although sizeable parts of the population may be suffering from this predicament, it has received little policy attention due to being confined in private, domestic spaces beyond the public gaze. However, domestic energy deprivation is differentially multiplied across larger spaces. McCormick and Philo (1995) have used the 'geography of poverty' concept to develop a comprehensive review of the multiple linkages between localities and deprivation in the UK. Among other texts, their study provided a key analytical foundation for Milbourne's (2004) examination of the multiple spatial, social, economic and cultural underpinnings of poverty in rural Wiltshire, whose conclusions underscore the 'importance of making sense of the different spatial scales of poverty, as well as the complex relationships that exist between poverty, people and place within particular spatial contexts' (p. 573). The post-socialist realm has yet to benefit from the empirical application of such ideas.

Theories: from relational space to evolutionary economics

The foundations of the conceptual framework of this book lie in the geographic literatures on economic, social and infrastructure reform in the developed world. The multiple connections between urban poverty, on the one hand, and broader dynamics of technological and institutional change, on the other, are of particular importance here, especially since they allow us to think about social inequality in relation to the emergent spatial divisions of labour and capital in post-industrial societies (see, for example, Leyshon and Thrift, 1995; Wills, 2004; Albeda and Withorn, 2002; Peck, 2001). Thus, Byrne (1999) points out that 'the spatial policies which matter in relation to social exclusion' have played a crucial role in the broader transition 'from an industrial to a post-industrial system' (p. 122). Theoretical discussions about the spatial implications of these dynamics have been facilitated by the profusion of new insights into the relationship between socio-economic change and poverty (see Room et al., 1989; Wacquant, 1999; Mingione, 1998).

Infrastructure networks have played a key role in the emergence of spatial differentiation among and within cities, by 'sustaining sociotechnical geometries of power and social or geographical biases in very real – but often very complex – ways' (Graham, 2000, p. 115). Sassen (1991), among other authors, stresses the permanent contradiction between the growing de-territorialization of global economic and cultural flows, on the one hand, and the spatial separation of certain groups and social classes from these levels of interaction, on the other. These trends are linked to the 'geographies of everyday life' (see Stenning, 2003; Holloway and Hubbard, 2000; de Certeau, 1984), which lie at the intersection of infrastructural transformations and social deprivation.

Aside from these general insights, however, I have drawn on a range of more specific approaches from human geography, planning, and welfare economics to develop a geographical interpretation of energy poverty in the given spatial context. A central idea in this context is the role of the home as an arena for different social and spatial dynamics, defined by networked relations among economic and social actors. In the post-socialist realm, the relationship between the domestic domain and social exclusion has been explored by, most notably, Hörschelmann and van Hoven (2003), who underline the limiting social and spatial role of the home in the case of middle-aged Eastern German women. I have also relied heavily on institutional and evolutionary systems of thought, which emphasize the path-dependency of decision-making processes in societies undergoing rapid social change. In the sections that follow, I explain the foundations of these theoretical approaches in further detail.

Understandings of poverty

The definition of energy poverty provided above may be placed in the framework of broader debates about poverty and deprivation *per se*. In this case, it is possible to distinguish several distinctive sets of ideas. Traditionally, economists have measured welfare using either a commodities- or a utility-based approach. The former originates from the Rawlsian theory of justice, which equates well-being with the possession of commodities and income. The utility approach draws its roots in Jeremy Bentham's eighteenth-century utilitarianism, whereby 'an action is right if it tends to promote happiness, and wrong if it tends to promote the reverse of happiness' (Bentham, 1996). The corresponding poverty-measuring method sees poverty as a function of 'utility', which can be interpreted as the fulfilment of a certain need, desire, happiness or choice. Both approaches operate with an income-based definition of poverty, in that it is assumed that the loss of either commodities or utility occurs below a certain amount of income, which can be absolute (as in a 'basic needs' subsistence level) or relative (where the most commonly-used principle states that a household is 'poor' if its income does not exceed 2/3 of the median income of the entire population).

In the 1980s, the hegemony of these two theories was disrupted by Amartya Sen, who proposed that the space of 'capabilities' is more appropriate for evaluating inequality (Sen, 1980). He argued that the possession of commodity or utility cannot provide proxies for well-being, but rather it is important to concentrate on what the person actually succeeds in doing with the commodity, given its circumstances. As a result, poverty is measured on the basis of variations in the individuals' 'capabilities to function', that is, 'their ability to have a long and healthy life, to be well-nourished, literate, safe, and so on' (Cornia et al., 1996: 161, also see Saith, 2001 for an operationalization of the capabilities approach). Amartya Sen's theory helped switch the emphasis of poverty analysis away from the 'means' (income, wealth and so on), and onto the 'ends' (quality and quantity of life), allowing for the incorporation of a wider range of factors. The UNDP's 1996 Human Development Report has concluded that 'while 900 million people in developing countries are income poor, 1.6 billion are capability poor' (UNDP, 1996, p. 2)

Having looked at the empirical evidence and the nature of all three theories, I concluded that the measurement of energy poverty requires a combination of all of them. This is because an energy poor family is simultaneously deprived of domestic energy (that is, the household is suffering from a loss of utility), due to falling real income or inadequate housing stocks (lack of commodities) and the individuals' decreased ability to keep warm (loss of capabilities).

A relational view of energy poverty

The book analyses energy poverty with the aid of a relational understanding of the interactions between society, economy and space. This approach stems from Graham and Healey's (1999) argument in favour of a 'non-Euclidean' relational understanding of contemporary urban social and territorial processes in planning practices, Pawley's (1997) discussion of 'terminal architectures', and Boyer's (1995) interpretation of social exclusion through the 'disfigured city', which 'has no form or easily discernable functions' because it is detached from 'well-designed nodes' (p. 82). Similarly, Amin (1999) talks of globalization 'a set of spaces of juxtaposed fragments and contrasts, where diverse relational webs might coalesce, interconnect or disconnect' (p. 43), while McDowell (2004) stresses the policy implications of the 'theorization of identity as relational'. Goss (1998) points out that 'space can no longer be conceived as merely material, nor social relations as merely abstract' (p. 402).

Relational definitions of deprivation and marginality have been present in the poverty literature for a considerable time, as a result of the mainstream acceptance of relative understandings of poverty (for a further discussion, see Townsend, 1979) and the theorization of social exclusion in terms of 'distributional and relational aspects', which are mutually interrelated, 'irrespective of the stage of economic development of a country' (Bhalla and Lapeyre, 1997: 422). Some key contributions to this field have placed social exclusion within the triangle of 'relativity, agency and dynamics', allowing it to be interpreted in terms of the multiple relationships between individuals and social processes (Atkinson and Hills, 1998). The notion of 'relational poverty' is gaining increasing prominence in the literature, as a number of international NGOs and development experts have used it as an explanatory tool for the interactions between participatory practices, institutional structures and social deprivation in the developing world (Tembo, 2003; Paugam, 1995; CDR, 1999). However, it has yet to translate into a tangible theoretical framework and policy praxis.

The relational approach used in this book operates on the premise that energy poverty – and poverty *per se* – can be understood both as a systemic process that lies at the intersection of economic, social and spatial policies (as outlined, for instance, by Yeung, 2005; Graham and Healy, 1999), and as a lived experience, arising from the mediation of everyday life through a household's social and/or built environment (McDowell, 2004; Boyer, 1995; Jarvis et al., 2001). These dynamics arise out of interconnected networks, which relate to each other through policy conflicts, structures of the built environment and the 'geographies' of everyday life. Through their fluid interconnections, the home becomes a 'prison', a space of virtual captivity that creates deprivation via its interaction with the households who use it. Thinking about the relationships between homes and households in this manner allows us to see the dynamic ways in which buildings and people 'are always already bound together, always already binding together' (Bingham, 1996: 635).

Some of the policy aspects of energy poverty have been interpreted through a strategic-relational framework, an approach formulated by Jessop (2001) and applied in practice by Jones (1997), on the basis of ideas initially proposed by Poulantzas (1973). Through it, the post-socialist state can be seen as 'strategically' and/or 'structurally' biased in the choice of transformation strategies, to the detriment of energy efficiency and social welfare. Thus, in strategic-relational terms, energy poverty is a product of the recursive mobilization of neoliberal strategies and projects, aimed at transforming the structural selectivities of the state. The wider application of this theory, I believe, can provide a first step towards the much-heeded re-conceptualization of economic and social approaches within poverty studies. However, it should be noted that the strategic-relational approach has a number of analytical weaknesses, such as the excessive focus on the national scale of capitalist production and the neglect of non-state agents and institutions.

Evolutionary approaches

Concepts borrowed from this burgeoning field have been used to analyse the divergent responses of post-socialist institutions to energy poverty-shaping processes (for a comprehensive discussion of the interface between evolutionary economics and economic geography, see Boschma and Lambooy, 1999; Reijnders, 1997; and Vromen, 1995). The application of evolutionary economics to the study of energy poverty implies that the economic and political history of a given locale shapes the institutional background and spatial structure of domestic energy deprivation. According to Dallago (1999),

> systemic convergence as an evolution of different mixed type does not exist ... economic systems either evolve or they disrupt ... systemic capital and transition costs together produce path-dependence as a rational response of individual actors attempting to reconvert their still-valuable old systemic capital and to economize on transition costs. Consequently, different economic systems may not converge simply because differences are reproduced, alongside adaptation, in evolving environments (p. 172).

This line of thought leads to the conclusion that economic institutions must develop organically and cannot be transplanted from one geographical context into another.

However, network analysts have argued that 'ambiguity can be a resource for economic action' (Grabher and Stark, 1998, p. 56). This resembles the conclusions reached by Dunford (1998) in his comparative analysis of the different modes of regulation and development in the post-socialist transformation. He argues that 'advocates of shock therapy did not anticipate the scale and durability of the collapse in output and income', and as a result, their formula included three principal failures: first, it did not anticipate the impact of structural adjustment programmes on economies that were not market economies; second, it failed to identify the nature of institutions on which capitalism depends; and third, it did not comprehend that modes of conduct, which are taken for granted in capitalist societies, have to be learned in the case of socialism (p. 108).

Extending the evolutionary argument to the case of different sets of organizations leads to the conclusion that selection inefficient organizations and arrangements may exist regardless of the external context (Nicita and Pagano, 2001). This finding highlights the role of 'path-dependency of the third kind': a situation whereby agents make inefficient choices due to cultural, ideological and/or opportunistic reasons, although they are aware of the alternatives (Magnusson and Otosson, 1997). Understanding the nature and origins of path-dependent decisions within post-socialist institutions is crucial to interpreting the broader background of scarcity and marginalization at the household level.

Institutional economics

Institutional economic insights have been employed in many instances throughout the book. In particular, the notion of 'transition costs' has been useful in explaining the emergence of path-dependency in transformation. 'Transition costs' can be understood as the price of transferring from one institutional mode to another (see Voigt and Engerer 2000, for a further discussion). Another frequently employed theorization is the 'institutional trap': an inefficient, stable and self-reinforcing institutional system, which exists due to 'norm stability', that is, the high costs of transition from one organizational norm to another. This idea stems from theories initially developed in Arthur (1988) and North (1990) and includes 'the co-ordination effect secured by a type of externality' (the more consistently a norm is observed in society, the greater the costs incurred by each individual deviating from it), as well as cultural inertia (which denotes the agents' reluctance to reconsider the behavioural stereotypes that have already proven viable). Thus:

> the hypothesis that efficient institutions must arise because of natural selection does not prove to be truthful … Each institutional transformation should be preceded by efforts to forecast and forestall possible institutional traps … A right choice of the rate and sequence of the reforms, and wise industrial policy, are prerequisites of institutional trap avoidance (Polterovich 1999: 121).

Institutional traps stem from poorly co-ordinated restructuring attempts in the early phases of the transformation process. This explains their high frequency in countries like Russia, Ukraine and other FSU states (although they can be found in virtually every post-socialist country). It has been established that the institutional trap in Russia is manifested in the form of non-monetized exchanges (particularly in the energy sector) and rent-seeking on the behalf of governing elites (Zaostrovtsev, 2000). In the relevant chapters of this book, such theoretical findings are used to analyse energy operations in post-socialist countries.

The institutional trap is linked to the broader 'under-reform trap' experienced by many FSU and some ECE states, wherein 'high corruption drives the economy underground, and the resulting low tax revenues make it hard to control corruption' (Åslund et al., 2002, p. 92). The 'under-reform trap' was first described in Johnson et al. (1997), while Hellman (1998) linked the theorization of the 'trap' to 'state capture', which is caused by entrepreneurs privatizing public goods, such as taxation and regulation to their own benefit.

Stucture of the book

The remainder of this book basically consists of two parts, the first of which (comprising Chapters 2 and 3) provides a general overview of the energy poverty problem, focusing on the background historical context for its emergence (in Chapter 2) and the situation on the ground (Chapter 3). Chapters 4, 5 and 6, in the second part of the book, are case studies of energy poverty in Macedonia and the Czech Republic, respectively. In a way, these chapters can be seen as a comparative study of the two countries, which represent different political-economic situations: Macedonia has more extensive 'geographies' of energy poverty, due to severe welfare shocks during the post-Communist transition, while the Czech Republic has been more successful in alleviating poverty, reforming the economy and achieving economic growth. There is, however, a great deal of asymmetry between the two case studies, because of the inconsistencies between source datasets and the different types of issues faced by each state. The respective analyses emphasize the most important problems faced by each country, rather than following a strict comparative structure.

The main objective of Chapter 2 is to investigate how the theory of post-socialist transformation has conceptualized the interdependence of energy pricing, poverty and demand-side energy efficiency in the countries of the former Eastern bloc. One of the main findings of the chapter is that the academic literature on the post-socialist transformation lacks sufficient theoretical knowledge about the interaction between state energy and social policies, the condition and usage of the housing stock, and the spatial extent of social deprivation. Energy poverty arises as a result of the poor co-ordination of such policies.

In Chapter 3 I examine the on-site consequences of energy, social and housing policies in post-socialism. The chapter looks at the patterns of poverty and income inequality, as well as energy reforms, efficiency and affordability across the post-socialist space. It provides a broad-level categorization of the outcomes of the transformation process in terms of the emergent geographies of energy poverty.

Chapter 4 looks at the institutional frameworks of the energy, social welfare and housing sectors in Macedonia and the Czech Republic. It aims to examine the relationship between energy poverty and broader governance frameworks: how does the emergence of inadequately heated homes relate to decision-making patterns in the relevant policy domains? The chapter pays particular attention to the various institutional 'lock-ins' in post-socialist economic reforms.

This is followed by Chapter 5, which undertakes a more detailed investigation of energy poverty evidence in Macedonia and the Czech Republic. It is based on data about aggregate household expenditure and the subjective perception of the level of achieved warmth in the home. Based on this investigation, the chapter constructs a socio-demographic profile of energy poverty in the two countries. One of its main findings is that the territorial and organizational contingencies of energy poverty are encased in a multi-layered geography of institutional traps, spatial inequalities and household practices.

Chapter 6 examines the everyday lives of households vulnerable to, or suffering from, domestic energy deprivation. It looks at the manner in which experiences of deprivation are shaped by residential occupancy and energy efficiency patterns. The interaction of fixed and mobile infrastructures in the residential energy sector is seen as a possible reason for the higher-than-average cost of energy for the poor.

In Chapter 7, I synthesize the reviewed empirical evidence, while using the findings of the previous chapters as a basis for discussing several broader theoretical and practical issues raised by the hidden geographies of energy poverty in post-socialism.

Chapter 2

Gaps in Theory and Policy: Tracing the Roots of Energy Poverty

For economic theorists the problems facing the socialist countries represent a challenge. Here we have a set of countries embarking on the choice of economic system. Surely economic theory should provide considerable guidance. Regrettably even science – at least until recently – has had very little to say about these fundamental matters, and even less to say about the important issues of transition (Stiglitz, 1994, p. 3).

Given that domestic energy deprivation in ECE emerged under specific social, economic and political circumstances, it is important to ask whether there are any special reasons for its rise in this particular context: what is it about the post-Communist transition that led to the expansion of domestic energy deprivation at this place and moment in time? In order to answer this question it is necessary to take a closer look not only at the developments that have marked the post-socialist transformation process, but also the concepts, knowledges and discourses that have guided decision-makers in the relevant countries. This chapter thus examines the gap between theory and practice in the production of energy poverty. It looks at the different ways in which experts and policy-makers have responded to the changing political and economic realities of the post-socialist world.

The key argument advanced here is that the growth of energy poverty in ECE and FSU was predicated by the inadequate and insufficient conceptual understanding of the driving forces of domestic energy deprivation. Although this 'knowledge gap' may be attributed, simply, to the unintended coming together of various circumstances that, in their entirety, are conducive to energy poverty, it can also be claimed that it is the product of a particular political economy, which consciously favoured certain policies over others. The aim of the chapter, therefore, is to examine how the existence of such a policy bias may have been grounded in the theory of post-socialist transformation, and the manner in which that theory was reproduced in the policy making process. In order to achieve this, it compares and contrasts a range of conceptual standpoints in the economic, energy, social welfare and housing literatures to the path of the reform process itself.

The chapter suggests that one of the reasons why energy poverty has become such a mounting problem in ECE and FSU is because its quick expansion – and potential political and economic weight – was not properly foreseen by energy policy theorists in the early days of the post-socialist transformation process. When the energy affordability issue did eventually come onto the policy and research map, it was often too late for concerted action to deal with its driving forces, with policy recommendations focussing on end-of-pipe solutions instead. Even today, most work in this field – scarce as it is anyway – displays an insufficiently inclusive conceptual understanding of energy poverty. This could, over the long term, lead to a greater accumulation of its social and spatial driving forces, while creating deep structural problems in the economy.

The chapter consists of four sections, the first of which provides an overview of the legacies of socialism at the onset of the transition process. This assessment of Communist-era energy, social welfare and housing policies serves as a basis for the second and third parts of the chapter, which investigate, respectively, the ideological struggles that shaped the post-socialist transformation process, and the conceptualization of energy and social welfare reform in the early years of post-socialism. The fourth section looks at how debates about the social aspects of energy reforms have been informed by the real-time results of theories that have already been applied in practice. It highlights the increasing awareness about the energy hardship faced by many East Europeans during the transformation process. The conclusion of the chapter critically revisits the multiple theoretical, discursive and decision-making incongruences that contributed to the emergence of energy poverty in post-socialism, while highlighting future research and policy challenges in this field.

The legacies of socialism

In order to understand the reasons for the emergence of energy poverty in post-socialism, it is necessary to look back at the policies and practices that defined Communist central planning. The socialist era was marked by a specific set of economic, social and political structures that had a deep influence on the regulation of energy, social welfare, housing and health.

The regulation of the energy sector: sidestepping the environment

Socialist planners were widely known for their heavy emphasis on energy security, supply expansion and industrial development (Gray, 1995, p. 1). Hence the immense intensity of energy production and consumption in Eastern Europe, which at the end of socialism had reached levels at least 50 per cent higher than those of comparable Western European countries (Hughes, 1991:79). According to Kramer (1991) 'The East European economies have developed, in the words of Hungarian economists, an "insatiable appetite" for fuels and power that engenders among them a "persisting propensity to overconsume" these resources' (p. 57).

The carbon-intensive fuel mix combined with the low efficiency of energy production, transmission and use, to exert unfavourable impacts on the environment (Meyers et al. 1994, Muiswinkel 1992). Kramer (1991) pointed out that the Communist planning model promoted 'policies that foster the excessive consumption of highly polluting fuels', which resulted in a 'massive degradation of the environment in many parts of the region' (p. 73). He emphasized the need for 'reductions in the overall consumption of energy, and in particular of solid fuels' as basic preconditions for improving environmental quality.

Most experts have attributed the 'energy intensity problem' to the nature of Communist economic ideologies, which placed a low value on energy and other natural resources, encouraging their overconsumption (Gray, 1995, p. 2). The socialist model 'viewed factor inputs, including fuels and energy, as inexhaustible and saw high levels of fuels and power consumption as indexes of modernization. The task of economic planners was to ensure that society's resources were mobilized in an all-out assault to achieve ever-higher economic goals' (Kramer, 1991, p. 59).

At the same time, the system promoted Russia's dominance in the supply of oil and gas to ECE and FSU. Partly as a result of these policies, consumers in other countries relied on low–quality, highly polluting oil 'wherever possible', because it reduced their dependence on Russian energy inputs and generated local employment (ibid.).

Not only was the cumulative fuel mix in the countries of ECE and FSU heavily skewed towards the most environmentally intensive and carbon-emitting energy sources, but the region's dependence on fossil (particularly solid) fuels was also heavily exaggerated for final energy consumption. Unlike Western Europe, where such forms of energy are usually concentrated on the supply side, the East inherited a socialist pattern of extensive utilization of solid fossil fuels by industry, services and households. Hence the high social and environmental significance of energy end use in ECE and FSU: 'This difference is critical in environmental terms … it is much more difficult or impossible to deal with the emissions generated by burning coal in small industrial boilers, heating plants, and domestic fires' (Hughes, 1991: 82).

The economic and political sustainability of such regulatory practices was warranted by the specific organization of energy operations within the planned economy. Under socialism, governments controlled and planned every aspect of the energy industry. Because they were rent-, rather than profit-seeking, enterprises did not have to face 'hard budget constraints', unlike their Western counterparts (Åslund, 1992; Åslund, 2002). Also, electricity utilities had their investment approved and financed by the state, often within the integrated spatial planning process (Stern and Davis, 1998: 430; Crnobrnja, 1991). The utilities themselves were 'vertically and horizontally integrated', which means that a single company would be responsible for the generation, transmission and distribution of any one given type of energy (electricity, gas, oil) throughout any given country.

Amalgamating energy pricing and social protection

The socialist system's poor environmental and economic record in the energy sector was a trade–off for its heavy emphasis on social welfare. The Communist planned economy treated a number of essential goods and services, such as housing, heating and health, as 'basic necessities', which were meant to be accessible to all. As a result, the consumption of state services by individuals and households 'was heavily subsidized … both housing rents and public transport fares were kept very low' (Duke and Grime, 1997: 885). This meant that the price structure for consumer goods bore 'no obvious relationship to the economic costs of production' (Stern and Davis, 1998: 430). The results of such practices were particularly evident in the energy sector, which was dominated by an indirect economic subsidy to consumers, financed by the state 'forgoing revenues as the owner of the assets' (ibid.). The result of this policy 'was a system of low energy prices in absolute terms (below international levels) and, in particular, low prices for households. Low household prices were sustained by relatively high prices charged to industry, the revenues of which were used to cross-subsidize household prices' (Gray, 1995: 1).

These implicit subsidies were motivated by the ideological principles of the political system, which saw social welfare as an integral part of the economic structure. Social policy in Communist societies was incorporated in all aspects of the economy, having also taken the form of price and industrial policy. Also, 'the core social programmes were largely financed by earmarked contributions of enterprises (payroll taxes), but funds were not separated from the general state budget and benefits not linked to previous contributions' (Götting, 1994: 183).

Neoclassical economists have argued that distorted pricing structures for energy carried the principal blame for the low efficiency of energy consumption and the poor financial viability of energy sector companies at the onset of the post-socialist transition. The inadequate ratio between residential and industrial energy tariffs failed to reflect the much higher relative costs of supplying energy to households, whose demand pattern is less predictable, and, unlike energy, cannot use the high-voltage grid:

> In most socialist countries, governments have sought to protect the interests of vulnerable groups by subsidizing the production of so–called necessities and by making a range of services available free of charge … Widespread subsidization of necessities is not only often inequitable, it is also inefficient. Subsidies distort relative prices. They give consumers misleading information about relative costs and thus lead to the misallocation of resources (McAuley, 1991: 100).

Housing and urban policies

Socialist central planning had an overwhelming effect on the planning, production and consumption of residential space in Communist societies. Nearly all ECE states were subjected to fundamental land reforms after the Communists came to power, resulting in 'widespread land ownership and small parcels' (Pichler-Milanović, 1994). Although an entirely new system was not introduced, some crucial modifications were applied in each country, thus changing the regulation of property management (Hegedüs and Tosics, 1994). Also, the state took control of nearly all aspects of housing operations, including constuction, purchases, sales – with most countries lacking an official housing market – and housing renovation. Many contries also had a nationalized rental sector, which was characterized by strict state control and the regulation of rents and the rights of tenants. Property rights were also subject to political control, although private ownership continued to exist in most states. This is evidenced by the 'one family – one flat' policy or the struggle against the 'luxury building': privately owned houses exceeding the common standard in terms of size and/or quality (Baross and Struyk, 1993; Enyedi, 1990; Szelényi, 1983).

The overwhelming interventionism practised by the socialist state had an equalizing effect on urban social segregation. While most socialist housing was substandard and modest, there were few demographic and spatial differences in the allocation of flats. It is stylized knowledge that members of markedly different social classes and professions would usually live next door to each other (Lux, 2000a). This is not to say, however, that there was no social segregation in Communist cities: according to Sailer-Fliege (1991), 'the specific socialist form of segregation was reflected by an over-representation of middle- and higher-status groups in socialist housing estates', as opposed to 'the general under-representation of these strata in the single-family housing sector in larger cities' (p. 11). She attributes this to 'the relative advantage of the different urban housing submarkets under socialist conditions', which interacted with 'the specific socialist form of housing allocation' (ibid.).

The land and rent management policies of socialist societies meant that low income households often had access to high quality housing in prime locations. This allowed housing capital to serve as an asset in the periods of economic hardship that followed the fall of Communism. Yet it was only in the mid-1990s that the 'real net value of housing' began to be incorporated in economic analyses of income distribution, with the aid of an imputed rental value (NIRV). Two separate studies by Buckley and Gurenko (1997) and Gillan (2000) concluded that the Gini coefficient of income inequality would decrease by 3 to 6 percent if the economic value of housing were to be taken into account:

> Surveys of living standards that fail to take into account the value of existing housing resources miss an important dimension of householders' coping strategies ... an estimated monetary value for that portion of housing for which residents were not paying the full economic cost could be obtained [by using the NIRV]. This represents an income-in-kind for households and is particularly important for less well-off households (Gillan, 1999: 260).

Residenial energy provision: energy efficiency issues

In their entirety, the specificities of socialist energy, social welfare and housing policies produced a unique pattern of household energy provision. Although the main types of energy sources used by ECE and FSU citizens did not differ significantly from those in other countries with similar climates and levels of development, they did possess a particular structure and distribution. In line with the Leninist credo 'Communism is Soviet power plus the electrification of the whole country', electricity was the most universally available source of energy, with most countries achieving impressive levels of network access, at more than 95 per cent of the population. Even though most urban areas in CEE and FSU also had access to piped gas, it was generally less available. This was especially true in low density or underdeveloped regions where the state wasn't willing to construct an entirely new, elaborate set of infrastructures (Rodionov, 1999; Arpaillange, 1995).

Most urban households in Communist countries obtained their domestic warmth via an extensive network of district heating (DH) systems. These structures used large, centrally-placed combined-heat-and-power (CHP) or heat-only-boiler (HOB) plants to burn coal, fuel oil or – less often – natural gas, in order to produce hot water, usually both for space heating and direct household use. The water was then transported through pipeline networks to substations, from where it was distributed to collective or individual buildings. In contrast to most Western European homes, where pipes are arranged horizontally – thus connecting all the radiators in any given apartment within a single loop – east European DH pipes in apartment buildings ran vertically, allowing radiators in rooms that were positioned on top of each to be supplied by the same pipe. As a result of this arrangement, the radiators in each home were connected to different pipes, making it more difficult to measure energy consumption at the household level. Adding to this issue was the general lack of metering and regulation for DH, which often resulted in too low or too high room temperatures and further losses of heat from the opening of windows to cool overheated rooms. Further efficiency problems in the network included hot water leakages due to both internal and external corrosion of pipes, as well as heat losses as a result of inadequate insulation (Kazekevicius, 1998; Lampiettu and Meyer, 2002).

Table 2.1   Share of households with access to district heating in selected post-socialist states

Country	Total
Lithuania	58%
Estonia	52%
Poland	52%
Belarus	50%
Czech Republic	32%
Romania	31%
Bulgaria	20%
Hungary	16%
Slovenia	15%
Serbia	13%
Croatia	9%

Source: Euroheat, 2006.

Even though DH requires high population densities in order to justify its high capital costs – which has made it an urban form of energy – socialist countries often constructed such infrastructures in smaller towns and even villages. This was particularly true in the former Soviet republics, which currently possess the highest shares of homes connected to DH, exceeding more than 50 per cent of all households (see Table 2.1). This percentage is the highest within urban areas: for example, nearly 90 per cent of city dwellers in Estonia have access to network heat, while the same figure reaches 80 per cent in Belarus. In Lithuania 'around 68 per cent of dwellings in towns are connected to the district heating systems', but as a result of Soviet legacies 'almost half of dwellings have no meters and their possibilities of heat control are very low' (Euroheat, 2006). Central European countries also possessed extensive DH systems: for example, 103 Hungarian towns currently have such networks. However, the share of this energy source in the final energy balances of Central European and Balkan states was significantly lower due to the presence of other fuels, such as piped gas and fuelwood.

Networked energy infrastructures were a basic component of socialist-era housing provision. They were one of the key reasons for the construction of large, centrally planned collective apartment buildings, which came to dominate urban landscapes in many FSU and CEE countries. Multi-family apartment buildings in FSU and ECE included a wide range of housing types, ranging from the luxurious Stalinist brick buildings of the 1950s – which are mainly located in inner urban areas – to prefabricated concrete panel housing estates placed at the outskirts of the city. Such blocks, which vary in height from three to more than 20 storeys, contained the majority of dwellings in urban areas: 73 per cent in Estonia, 50 per cent in Lithuania, and approximately 80 per cent in the Russian Federation (Lampietti and Meyer, 2003). Socialist residential buildings suffered from above-average heat losses due to a combination of factors, including: low thermal efficiency standards, a historical lack of attention to quality in construction materials and practices, and inadequate levels of maintenance. Common problems would include leaky windows and doors, uneven heat supply within buildings, as well as missing or insufficient basement and roof insulation. In many prefabricated panel buildings, the rubber moulding and cement mortar between panels quickly deteriorated, permitting air and rain to pass through (ibid.).

This is just a selection of the structural issues faced by Communist states at the onset of the post-socialist transformation. In their entirety, they posed a complex policy challenge that also required deep economic and political changes in the fabric of society. It is for this reason that the next section 'zooms out' onto the broader context, to look at the main ideologies of economic transition *per se*.

The end of Communism: initial debates

When Communism collapsed at the end of the 1990s, the future reform path towards a market-based economic regulation became highly controversial. Policy-makers and experts across the region – and beyond – had major disagreements about the manner in which former socialist countries should restructure their economies and societies. This conflict is typified by the controversies surrounding the term 'transition', which was – and still is – widely used to describe the transformation process. It was coined by neoliberal economists and politicians to indicate that ECE and FSU states were restructuring their economies in a unidirectional manner: from a command economy to a market-based system. But others, including those who argued for a more gradual and open approach towards the reform process, criticized these ideas for being politically teleological and conceptually limited (Stenning, 2005; Pickvance, 1997; Wälde and von Hirschhausen, 2001). It was claimed that the notion of 'transition' is 'neglectful of the embeddedness of the diversified transformations taking place' (Smith, 1997; Stark, 1992).

The 'transition' debate was at the tip of a much greater iceberg, which included the deep seated ideological and conceptual differences between 'shock-therapists' and neoclassical economists, on the one hand, and regulationists, institutionalists, 'gradualists' and development theorists, on the other. These debates concerned the normative speed, extent and content of the transformation process, which remain a contentious issue to date. Although many such discussions are already well known and publicized, it is worth revisiting them here briefly, in order to set the scene for the discussion of energy and social welfare reforms that follows.

The early triumph of neo-liberalism

The initial years of the transformation were marked by a political victory of the neoliberal regulation over rival policies and ideologies. Most countries pledged to apply a 'Washington consensus' -based approach in restructuring their economies, in spite of the numerous conflicts and controversies between the different theorizations of post-socialist transformation. This prescription was soon integrated into the policies of international financial institutions and the theoretical template of 'transition economics', although, 'a complete tableau of transition economics was never fully specified, and precisely what policy instruments were supposed to achieve was not made absolutely clear' (Amsden et al., 1994:16).

The neoliberal (or 'market fundamentalist') argument was further strengthened by the proponents of the 'shock therapy' model, which derived its name from Poland's stabilization and liberalization programme, initiated in 1990 (Marangos, 2002). This approach was followed by Czechoslovakia (starting from January 1991), Bulgaria (February 1991), Russia (February 1992), Albania (July 1992), Estonia (September 1992) and Latvia (June 1993). Its main proponents were well known economists based in Western countries, including Anders Åslund and Jeffrey Sachs. The latter was an advisor to the Polish and Russian governments; both he and Åslund guided the Russian reform process between 1992 and 1993 (Schlack 1996). Sachs and Aslund 'shared the belief that the economy [in Russia] was in such a terrible mess that a radical, comprehensive, liberal program would be needed to introduce any kind of rational order' (Schlack, 1996: 16, cited in Marangos, 2002).

While a detailed discussion of the theoretical content of 'shock therapy' extends beyond the remit of this book, it is worth noting that its mainstay was the idea of rapid economic change, in order to take advantage of an 'extraordinary' period of politics, which offered a 'window of opportunity' to simultaneously move towards both a market economy and a democratic political system (Åslund, 1994: 65). This would be achieved by abolishing most 'regulations on enterprise, trade, prices and production' (ibid., p. 67) in addition to rapid macroeconomic liberalization and financial stabilization. Progress along such a reform front was to be measured via five benchmarks of economic performance: inflation, national income, consumption, unemployment and income distribution. The same group of policies was also expected to help address rent-seeking and profiteering, leading to more competition and a 'healthier' distribution of income.

One of the reasons for the initial success of 'market fundamentalists' lay in their ability to persuade the public that a broad theoretical consensus existed with regard to market reform in former socialist countries (see Summers, 1992: 7). For example, Sachs argued that:

> The first, basic steps to the transformation of Eastern Europe's centrally planned economies are two. One, the Eastern countries must reject any lingering ideas about a 'third way' such as chimerical 'market socialism' based on public ownership or worker self-management, and go straight for a western-style market economy. Two, Western Europe, for its part, must be ready and eager to work with them, providing debt relief and finance for restructuring, to bring their reformed economies in as part of a European market system (Sachs, 1990: 22).

It seems that a combination of politics, power and ideology – rather than economic efficiency arguments *per se* – lay behind the initial prevalence of the line of thought advocated by economists such as Sachs. The neoliberal paradigm was endorsed by the world's most powerful international financial bodies, notably the Bretton Woods institutions (the World Bank and the International Monetary Fund). These organizations served as promoting agents of liberalizing policies long before 1990, having adopted a neoclassical development approach in the early 1980s (Sidaway and Pryke, 2000; Tucker, 1999).

Gradualism and institutionalism

Opponents of the neoliberal formula gained sufficient political strength only after 'shock therapy' was introduced in practice, and many ECE and FSU states began to face a new set of economic and social problems (UNDP, 1998; Elman, 1997). The theoretical foundations of neoliberalism were challenged by the emergence of 'a more social democratic way of thinking' (Lavigne, 1999, p. 16), which stressed the importance of sequencing the reform process in line with antecedent cultural and economic layers (Fischer and Gelb, 1991). Despite opposing this approach, also termed 'gradualism', Åslund provided a succinct overview of its key tenets in one of his books:

> Arguments against a quick transformation tend to emphasize the dangers of massive unemployment and to point out that restructuring is a process that takes time. In particular, a gradual deregulation of foreign trade has been proposed ... Another argument for gradual change is that social costs will become excessive. Such an assertion suggests there is an alternative approach that will cause less suffering ... A third argument against swift transformation is that the economic nadir is likely to be more profound under shock therapy and that such a point will be so low that a social explosion will ensue (Åslund, 1992, p. 34).

The gradualist case was supported by the 'institutionalists', who asserted that markets can 'work' only when the necessary institutional framework is established. They emphasized the centrality of 'institution building' in the post-socialist transformation process, while emphasizing the importance for further research in this field (Raiser et al., 2000). Such theorists also stressed the importance of introducing a clear regulatory framework for facilitating investment, trade, competition and technology transfer. For instance, Amsden et al. (1996) argued that the power of institutions to shape 'the form, substance, direction, and pace of economic expansion' represented a critical difference between the classic capitalism that the Russians and Eastern Europeans have tried to promote, and the late industrial capitalism that the East Asians have pioneered. This is because the role of institutions in Eastern Europe was minimal in the post-socialist period, when resource allocation was left almost entirely to undirected and inefficient market mechanisms (ibid.).

Institutionalists also linked the effectiveness of economic transformation to the establishment of an adequate regulatory and organizational framework. Cornia (1994) pointed out that the success of price liberalization, exchange rate and foreign trade reform and privatization is contingent on the 'existence of adequate "institutions" guaranteeing clear and enforceable property rights (of whatever nature they might be), free entry and competition, correction for 'externalities' or other market failures, the establishment of social safety nets and the financing of the provision of these "public goods" through taxation' (p. 574).

The institutionalist argument was soon incorporated into a broader range of theories, which challenged the conventional neoliberal view that 'transition is a relatively unproblematic set of policies involving economic liberalization and marketization alongside democratization, enabling the creation of a market economy and a neoliberal polity' (Smith and Pickles, 1998, p. 1). According to Smith and Pickles, neoclassical transition economics 'reduces the complexity of the political economy of post-Communism', because shock therapy is unable to move beyond a prescribed set of instrumentalized recommendations. They underlined that even in Eastern Europe's 'most liberal polity', the implementation of economic best practice has been hampered by the failure to consider the vital role of institutional networks in social and political reforms.

Social welfare and the restructuring of energy industries in post-socialism

The tension between neoliberal and 'market fundamentalist' approaches, on the one hand, and institutionalists and gradualists, on the other, was reflected in the reform of energy and social policies in the post-socialist transition. Most countries decided to adopt a neoliberal energy reform strategy, in line with neoliberal principles which advocated the quick liberalization of the energy sector, accompanied by the unbundling and eventual privatization of state-owned, vertically- and horizontally-integrated electricity, gas and petroleum monopolies. However, such approaches were critiqued by gradualists and institutionalists, who argued that 'fast-track' restructuring approaches paid insufficient attention to the complex organizational and legal preconditions for post-socialist energy reform (see, for example, von Hirschhausen and Wälde, 2001). It was unclear how a move to the brutality of the free market could be implemented in conditions of widespread poverty and political instability. According to Velody et al. (2003) 'Preparation for privatization and partial privatization has led to continued calls by international financial organizations for increased tariffs, but consumers in post-Communist states have responded skeptically, claiming that utility bills are already too high' (p. 24).

These debates pointed to the importance of sequencing economic and regulatory reforms in post-Communism. It became evident that the choice between 'market fundamentalism' and 'institutionalism' is dependent upon the political feasibility of each strategy, as the former was more likely to be adopted in a situation of economic crisis, when the political support for radical economic measures was there. While the former was more likely, the latter was promoted in conditions of economic stability, when the support for radical action was subdued (Machonin, 2000; Sojka, 2000; Kreidl, 1999).

Neoliberal energy reform frameworks

Early 1990s normative discourses about post-socialist energy policy were dominated by the neoliberal paradigm of economic transformation. The main principles of energy industry restructuring were outlined by Stern and Davis (1998):

> First, it is expected that 'the main consumer groups should pay the full economic costs of the production, distribution and supply of the electricity that they consume' (ibid.). Second, an effective transition must provide for 'the establishment of commercially viable electricity companies that can finance their investment requirements without recourse to subsidies or other financial assistance from the state (p. 428).

The pricing argument was further refined by Dale Gray, Principal Economist in the World Bank's Private Sector Development Department. He argued that 'the level of prices should allow utilities enough to cover operating cost, debt service, to finance a significant portion of investments from internal cash generation, and to remunerate assets through payment of adequate dividends' (Gray, 1995, p. 24). However, 'the structure of prices should be based on indications derived from LRMC [Long run marginal cost] estimates' (ibid.).

One of the neoliberals' main arguments for fast-track reform and deregulation post-1990 was that the distorted energy pricing structure inherited from socialism prevented energy utilities from raising the necessary investment revenue (UNECE, 1991; Muiswinkel, 1992). This was also applied to the case of efficiency: most transition economists though that rationalized energy prices would lad to more efficient energy use. As a result, the neoliberal reform approach centred on the promotion of 'getting the prices right' in a liberalized and 'marketized' environment, in order to deliver 'allocative, distributional and dynamic efficiency gains' (Gray, 1995, p. 39). It was suggested that more efficient pricing would easily lead consumers towards more economically-efficient decisions *vis-à-vis* energy efficiency, because 'energy efficiency improvements are theoretically profitable on their own' (Strickland and Sturm, 1998: 877).

Kramer (1991) argued that post-socialist countries should 'increase prices for energy resources to reflect their real marginal cost and thereby provide consumers with a positive economic incentive to conserve energy' (p. 14). Theorists like him predicted that energy tariff rebalancing will bring faster-than-expected results, if implemented as quickly as possible. According to Hughes (1991), 'the target of reducing the overall energy intensity of GDP by 50 per cent for Bulgaria, Czechoslovakia, and Poland, and by 30 per cent for Hungary within 10 years is a goal that can be achieved by a combination of pricing policies and industrial restructuring to improve or eliminate the most wasteful industrial activities' (p. 96).

The pricing argument was part of a broader effort to move the energy sectors of post-socialist economies towards a market-based regulatory system, similar to the one implemented in the UK under the Thatcher government in the 1980s, and later followed by Australia, New Zealand, Canada, several US states, South America and the Scandinavian countries. This regulation aimed to introduce private capital and competition by creating – wherever technically possible – new independent companies in the domains of energy generation and distribution. They were to face competition from each other and from new energy production companies. At the same time,

> transportation – where still necessarily a natural monopoly – is opened up by third-party access of competitors to electricity grids or gas pipelines which are owned either by private companies or by publicly owned companies exercising only a transportation function ... Under the influence of independent regulatory agencies, competition develops or is promoted by the regulators and sometimes by competition authorities (Wälde and Hirschhausen, 1998: 14).

This model was expected to lead to competition 'on the level of retail electricity and gas distribution as well as power generation and gas extraction and trading' (ibid.). Having been reinforced by international frameworks such as the EU Electricity Liberalization Directive, it came to dominate the path of energy transformation followed by most post-socialist states. It was also included in the reform discourses of influential multilateral banks active in the region; through various forms of loan conditionality, their policies had the power to drive energy policy transformation. Thus, the official policy standpoint of the European Bank for Reconstruction and Development (EBRD) was that 'in general, the Bank supports a reform path comprising (1) corporatization and commercialization of public sector entities, (2) demonopolization and (3) privatization ...' (EBRD 1999, p. 10). Moreover, 'energy enterprises should operate on commercial principles', which can only be effective if the operational environment is (among other requirements) characterized by: 'cost recovery and cash collection to levels that enable the supplier to be financially viable' and 'effective structuring of subsidies for low-income users of heat' (EBRD 1999: Annex 9, p. 1).

Post-socialist poverty: equity or efficiency?

> It is unfortunate that the collapse of the iron curtain occurred at a time when Thatcherite policies held sway in the West, resulting in such remedies being recommended for Eastern Europe. The recent change in the political climate in Western Europe may encourage a re-think on the nature of the transition and the future direction of Post-Communist societies (Duke and Grime, 1994: 889).

The proponents of gradual, institutionally-based economic reforms in transition argued that 'shock therapy' led to a quick rise in poverty and income inequality (see, for example Ellman, 2001). They pointed out that 'the social cost of reforms has been very high' (Kuddo, 1995: 65), while blaming neoliberal policies for the emergence of various types of social exclusion in the transition economies. According to Kumssa and Jones (1999), 'most of the former socialist countries, which had achieved impressive social development in the past, are now facing problems of poverty, malnutrition, homelessness, ill health, crime, as well as rising income and social inequality' (p. 12).

However, these claims were often critiqued for failing to explain why economic reforms should be concerned with distributional outcomes, provided that their efficiency over a certain time scale is not compromised by political unrest stemming from deprivation and poverty. Many neoliberal economists and 'shock therapists' argued that social justice is irrelevant to economic policy, as long as there is sufficient short-term popular support for liberalization, which would benefit the entire population over the long run. But even within the neoliberal camp there were different views as to how economic efficiency should be reconciled with social equity. Atkinson and Micklewright (1991) classified these theoretical standpoints into three categories:

- One line of thought claimed that 'the introduction of a market economy is justification in itself of the reform', and to the extent that this leads to inequality, it has no ethical consequences. Income inequality is 'justifiable' because it is 'generated by a process of free exchange'. Provided that the initial point of departure is regarded as 'fair', then 'no question need be asked regarding the outcome' (p. 5).
- Second, there were viewpoints which accepted that 'distribution may be of concern', but not without noting that 'inequality is an inevitable consequence of moving to a market economy and that there exist no effective policies to redistribute income ... consistent with individual liberties' (ibid.).
- A third perspective argued that redistribution, while possibly desirable in principle, has 'too high a cost in terms of economic efficiency'. As a consequence, the interests of all are best served by giving priority to 'improving overall economic performance ... In a growing economy, the gains from higher GDP per capita will "trickle down" to the lowest income groups, so that all will benefit in real terms' (p. 11).

However, gradualist and institutionalist supporters rejected all such perspectives because of their reliance on a particular set of conjectures, such as the idea that the 'initial point of departure was fair'. In this context, it is worth noting Smith's (1992) critique of 'the common assumption that the restructuring of socialist society inevitably involves the adoption of a "market economy," accompanied by at least a partial reversion to private property relations' (p. 129). He maintained that the only 'departure of such assumptions is from real life itself':

If the efficiency benefits to be derived from market mechanisms are to be associated with distributive outcomes with some defensible claim to equity, then the issue of social justice must come first. Dependence on the pre-existing distribution of income, wealth and resources risks perpetuating past inequalities if privatization means that the ownership or control of state assets is simply to be transferred to those who can afford them (p. 137).

Smith furthermore, argued that 'it is axiomatic that the distribution of income and product generated by market forces depends on the preexisting distribution. Very simply, those with most money (and other forms of property) exercise greatest influence on market outcomes. Thus, an unjust distribution (by whatever criteria) has an element of self-perpetuation, and with it other problems of the past' (ibid.).

Such critiques of the market fundamentalist model were extended to the wider dynamics of globalization in ECE and FSU, because, 'rather than eliminating or reducing differences, the integration of national economies into a global system has on the contrary made those differences more apparent, and in many ways, more unacceptable' (ILO, 1995, p. 6). This is in line with Bhalla and Lapeyre's (1999) opinion that 'there is a dramatic paradox between the mainstream discourse on the net benefits of greater integration into the global economy ... and the reality of the rising spiral of human insecurity for the bulk of the world's population' (p. 104).

At the same time, social policy experts attacked the 'total welfare abstinence' position on both equity and efficiency grounds. It was pointed out that 'if the majority of the population does not receive the protection and assistance that it needs, the transformation process is likely to be accompanied by significant political strife, and the switch to a market economy could be frustrated' (Schöpflin 1991). Götting (1994) doubted the 'purported trickle-down effect that the poor, too, will benefit from the growth dividend achieved' (p. 1983). She thought that this would be unlikely to happen without re-distributive social policies designed specifically for that purpose. Some authors claimed that the unequal distributional outcomes of 'shock therapy' could also compromise economic growth, because 'social degradation can destroy existing societal practices and institutions, and impoverish the people. From a purely economic view, policy which neglects the common welfare will adversely affect the stock of human capital and reduce productivity' (Kumssa and Jones 1999: 203).

Although the equity vs. efficiency debate has remained unresolved to date, it had important policy implications across the post-socialist space. Considering that the neoliberal paradigm of economic transformation did not give precedence to issues of social welfare and deprivation, the question arises as to how and whether this theoretical viewpoint was translated into the neoliberal regulation of energy reforms in transition.

The treatment of energy price subsidies and social welfare within the neoliberal paradigm

It is one thing for the coming of the market economy to bring lower cost cars, televisions, clothing and reasonably affordable, good quality foreign goods. It is quite a different thing for it to bring threefold, fourfold or more increases in rents, heating charges, electricity and gas prices, rail and bus fares, telephone and postal charges, water and sewage charges, health fees, etc. There is widespread opposition to the upward rebalancing of the prices of these commodities (Niggle 1997: 326).

In spite of the problematic ethical consequences of creating a more unequal income distribution as a result of 'shock therapy', the social aspects of energy reforms deeply affected the progress of transition efforts. This is because, as can be deduced from Niggle's opinion cited above, the removal of energy price subsidies created political tensions that undermined the speed and direction of the restructuring process.

However, neoliberal energy policies in Eastern Europe were poorly informed by on-site research of the social welfare implications of energy price reform. Normative arguments about the distributional consequences of 'shock therapy' in the energy sector did not have an investigational basis. Only two major empirical studies of this issue were published until 2001, both being undertaken *post festum* of the neoliberal recommendation that rapid price liberalization should constitute the first step of energy restructuring.

Thus, for a long time, the only refereed article on the issue was Freund and Wallich's (1996) work on *The Welfare Effects of Raising Household Energy Prices in Poland*. Their empirical analysis was based on data from the 1993 household budget survey, 'which contains information on the expenditures of 16,044 Polish households, surveyed between January and June 1993' (Freund and Wallich, 1996: 55). Examining the expenditure patterns of households in five equivalent income quintiles led the authors to conclude that 'not only did the better off spend a larger absolute amount on energy than the poor, they also consumed a larger proportion of their expenditures as energy' (ibid.). The subsequent conclusion that 'as income increases, people spend a larger portion of their income on energy', was used as a starting point for assuming the price elasticities of energy demand among different income quintiles.

Freund and Wallich (1996) calculated the loss of consumer surplus as a percentage of total expenditure among several social and demographic groups with different income elasticities, following an 80 per cent aggregate energy price rise. According to them 'Assuming a zero elasticity of demand, the poor's welfare declines by 5.9 per cent, while that of the richest quintile declines by 8.2 per cent ... For all consumers taken together, the welfare loss ... is between 4.6 per cent and 7.6 per cent of their total budget, depending on the assumed price elasticity of demand' (Freund and Wallich, 1996: 58). Their study also examined the welfare effects of energy price increases on different socio-economic groups. It emerged that 'farmers and mixed families are hurt the least ... largely because they do not consume district heat' (ibid.). Still, a significant decrease in pensioners' and workers' energy expenditure was expected to occur, due in part to the fact that urban pensioners tend to be 'house rich and income poor', especially in urban areas (p. 46). Thus, heating and lighting their relatively large dwellings placed a higher burden on family budgets.

Freund and Wallich also emphasized that 'the welfare loss to all consumers of raising electricity prices is much higher than other fuels, given its generally larger share in overall consumption'. Raising power prices by 80 per cent would result in a welfare loss of 2.6 per cent of aggregate expenditure among all households (as opposed to 1.7 per cent for district heat and 1.5 per cent for gas), with an 'especially notable' impact on the poor, whose welfare losses would reach 2.9 per cent, compared to 1.2 per cent for heat and 1.1 for gas. This prompted the authors to propose the introduction of 'special transitional approaches and pricing mechanisms to offset the social costs to low income households', including 'lifeline rates, vouchers, increasing social assistance payments, and general adjustments to wages and pensions' (ibid., p. 62).

The uniqueness of this study was ended in May 2001 with the publication of a working paper titled 'How much do electricity tariff increases in Ukraine hurt the poor?', by Dodonov et al. (2001; a revised version was published in the journal *Energy Policy* three years later as Dodonov et al., 2004). Despite the clear aim expressed in the title, nearly half of this paper was devoted to describing 'current pricing approaches for households in Ukraine'. The remaining sections of the study dealt with the poverty effects of a cost-covering tariff (estimated to range between 15 and 55 per cent higher than the current one in the short term, and up to 314 per cent over the long run). They used the compensating and equivalent variation to calculate the changes of well-being incurred by the needed price rise, in a nationwide sample of 126 representative households, disaggregated by income quintiles – according to aggregate income. It emerged that the energy burden of the three bottom quintiles would increase by at least 1 per cent, in the case of the lower short-term price increases. However, the energy burdens of the bottom income quintile would reach 44.92 per cent if they were to maintain their energy consumption, in a future situation with long-term, OECD-level energy prices.

Both studies received positive feedback in academic and policy circles alike, with Freund and Wallich 1996 being cited widely in the literature on energy sector reform. With the exception of Bacon (1995), who critiqued their use of elasticities and consumer surplus to estimate social welfare in conditions of 'very high price increases', most references to Freund and Wallich (1996) gave credence and authority to their findings. In particular, there was a focus on the conclusion that 'energy price subsidies benefit the rich more than the poor', which became stylized knowledge in policy discourses on energy transformation, having been extended to the entire post-socialist context (for example see EBRD, 1999, p. 13), although the data in Freund and Wallich (1996) and Dodonov et al. (2001) referred only to Poland and Ukraine, respectively.

The extensive reliance on the empirical findings of these two studies is theoretically problematic. Although the generic limitations of the analytical tools used by them were widely known in the broader literature on poverty measurement, it seems that they were not taken into account when citing the results of these studies. Three issues are of particular importance in this regard:

- First, both studies relied exclusively on household budget survey data. It is well-established that Eastern European household expenditure surveys are riddled with methodological limitations, due to, for instance, widespread under-reporting of income, poor disaggregation of expenditure data, and inaccurate census results. This may be a particular problem in Dodonov et al. 2001, because their empirical basis consisted of 126 households that were meant to be representative for the whole of Ukraine. Although the authors did try to show that their sample is statistically significant in terms of appliance ownership and demographic composition, the framing of their paper's empirical basis within broader social structures in Ukraine was hampered by the poor state of household socio-economic data in that country.

- Second, the underlying assumptions of both papers were questionable. The statement that 'higher-income families spend a greater portion of their income on energy' may be critiqued on the basis of Freund and Wallich's (1996) classification of households into income quintiles according to equivalent earnings on the OECD scale. As a result of this division, pensioners (who have the highest energy burdens, and higher-than-average equivalent incomes) would be clustered in the higher income quintiles, which may explain why 'the rich' appear to be spending more than 'the poor' on energy, in relative and absolute terms alike. This is despite the fact that Freund and Wallich themselves stated that pensioners can not be considered 'rich' in terms of their income circumstances, and indeed it was concluded that they would suffer the most from the removal of universal energy price subsidies.

- Third, both studies equated energy expenditure to energy consumption, assuming a one to one relationship between the two. This is in direct opposition to the broader literature on energy poverty, which stresses that the actual level of warmth and lighting in the home may not necessarily be correlated to monetary expenditure (see Boardman, 1991). The energy efficiency gap (that is, the difference between the delivered technical energy and the actual domestic energy service) has a distorting effect on the expenditure-to-consumption ratio, so that low-income households in badly-maintained homes receive little domestic warmth, despite paying a high monetary cost for the 'raw' energy arriving at the outer limits of the dwelling (Sanstad and Howarth 1994). In addition, aggregate energy expenditure comprises both unit costs and standing charges, which are normally of a fixed character and do not correspond to the household's energy consumption. Absolute energy expenditure data overestimate the energy consumption of families with high energy bills, because standing charges constitute a smaller proportion of their total energy expenditure compared to low-energy use households.

One aspect of these two studies, however, is particularly relevant to the theorization of energy and social welfare in transition: both of them promoted the separation of universal price subsidies from energy tariffs and the creation of new social safety nets to help the needy. This echoed Gray's (1995) earlier recommendation that 'significant price increases for residential consumers should be undertaken as quickly as possible, along with mechanisms to moderate the impact on low-income consumers' (p. 58). The implicit assumption within such arguments is that the distributional effects of energy reforms should be addressed by other policies and government departments. It was expected that social welfare and housing institutions will have the capacity to provide new forms of social protection in lieu of energy price subsidies. Was this, indeed, the case?

In the meantime: social welfare in transition

One of the facilitating aspects of the design of social policy in post-Communism was that policy makers did not have to 'start from scratch ... and create new institutions', as it was often sufficient to 'rebuild the already existing welfare state institutions step by step and adapt them to the environmental conditions of a market economy' (Götting, 1994: 182). However,

> In the transition period, political actors cannot only concentrate on 'normal' distributive policies, but also have to decide on basic questions of institutional design. The transition from plan to market implies that they must draw an institutional distinction between economic and social policies, specify the duties and revenues of the corporate actors who are to carry out social policy functions, decide on the public-private mix of social benefit provisions, and provide for what now has become a 'standard risk' of the production process, namely the risk of unemployment (ibid.).

Social policy in transformation was assigned an ever-increasing role, which included the protection of 'vulnerable groups, those on fixed incomes and those whose attachment to the labour force is marginal at best', as well as the insurance that 'these groups are compensated for both inflation and the loss of subsidies' (McAuley, 1991: 101). Moreover, rapidly rising levels of inequality often required the adoption of a general programme of anti-poverty measures (ibid.), in a situation where the state also had to create a favourable environment for further development by 'lowering taxes, limiting redistribution measures, and economising manpower [sic] and other resources for social policy' (Kuddo, 1995: 65).

Such tasks had to be managed by under-staffed, under-financed and untransparent institutions. The rebuilding of social safety nets in ECE and FSU was often the business of a few technical experts, who worked behind closed doors (Götting, 1994). Their work was further complicated by continuing fiscal crises, pressing social needs, and political instability (see Sykes et al., 2001; Ferge, 1992). According to Götting, constitutional decisions about social policies were 'often blended with (and sometimes hard to distinguish from) emergency policies merely intended as stop-gaps to alleviate the 'social costs' of societal transformation' (Götting, 1994: 182).

Given their complex policy tasks and stretched technical resources, it is easy to see why social welfare decision-makers neglected the creation of new social safety nets for energy price increases. Even if the political and technical capacity for such policies was in place, their knowledge basis was shaky. While the experience of post-socialist countries in restructruring their energy, social welfare and housing sectors was unique in history, it was not met with any theoretical or policy research based in the region: during the 1990s, not a single academic article, policy document or research report provided the region's governments with context-sensitive guidance about the design of social support mechanisms to improve energy affordability. This discrepancy reveals a gap between the regulations of energy and social reform in post-socialism. Energy theorists did not anticipate that social policy-makers in ECE and FSU would be unable to carry the added task of providing a new social safety net for the upward rebalancing of energy prices.

Evolving theoretical and policy responses to new socio-economic realities

The mid- to late-1990s saw the gradual incorporation of institutional- and political economy-oriented approaches into the mainstream economic reform agenda. A key thrust in this direction was provided by institutionalist work on the 'transformation crisis', which attributed the rapid fall in production and social anarchy in the FSU to the collapse of the state and the adverse socio-economic consequences of 'shock therapy' (Schmieding, 1993; Ellman, 1994). This body of research stressed the distinction between 'formal' and 'non-formal' regulation, based on the understanding that institutions 'encompass not only bureaucracies and administrations but also, and more importantly, the entire body of formal laws, rules and regulations as well as the informal conventions and patterns of behaviour that constitute the non-budget constraints under which economic agents can pursue their own individual ends' (Schmieding, 1993, p. 233).

Such ideas were also used to challenge the dominant view of energy transformation, which, as noted earlier in this chapter, involved far-reaching programmes of unbundling horizontally- and vertically-integrated monopolies, opening up energy markets to third parties and the privatization of energy utilities. Although the demise of socialist central planning in the early 1990s was followed by widespread political enthusiasm about the universal applicability of this Anglo-Saxon regulatory model, consecutive energy sector reform efforts progressed slower than anticipated in virtually all post-socialist states. This is despite their different starting positions and the varying degrees of international pressure and assistance (Stern and Davis, 1998: 428; also see von Hirschhausen and Opitz, 2000; Braber and van Tongeren, 1996).

The slow pace of the restructuring process was attributed to a combination of economic and political factors, such as institutional frictions, regulatory incompatibilities and under-reform traps. It was particularly pronounced in efforts to privatize energy companies. Considering that investors in Eastern European countries were constrained not only by financial, but also by political, macroeconomic and regulatory risks, the privatization process mainly took the form of 'sale of shares in state-owned energy suppliers or of their assets such as generation plants, which carries long-term investment and management commitments for the private shareholders; and ... concessions to invest in existing or new supply capacity to attract limited or non-recourse financing' (EBRD, 1999: Annex 9, p. 5). In the countries with weak formal legal cultures 'privatization' and 'deregulation' effectively amounted to a superficial reorganization of the sector, allowing real control to remain in the hands of governing elites (ibid.; and Opitz, 2000).

The real-life progress of post-socialist energy reforms vindicated the institutionalist argument that a 'legal or a professional lawyers' culture' (Wälde and Hirschhausen 1998, p. 9), accompanied by an appropriate institutional framework, is imperative for the establishment of 'independent, profit-oriented enterprises, acting according to 'capitalist criteria in a monetarized environment' (ibid.). It became clear that the failure to adopt a comprehensive legal framework for utility reform led to the uncontrolled emergence of domestically created *sui generis* solutions, which hindered the development of socially- and economically-effective institutions (Jasinski and Pfaffenberger, 2000; Newbery, 1994). The difficult transformation of the energy sector pointed to the need for developing comprehensive solutions for structural reform, in harmony with the legal, cultural and social conditions of the post-socialist context. It highlighted the multiple linkages between energy commercialization, unbundling and privatization, on one hand, and social welfare and energy efficiency, on the other.

One of the key contributions to the institutionalist argument for energy reform has been provided by Ürge-Vorsatz et al.'s (2006) overview of the 'policy agenda towards a sustainable restructuring of the energy sector in Central and East European countries'. Having looked at the history of energy policies in the region – which is riddled with environmental pollution and immense energy intensities – the authors point out that 'radical economic reforms and energy sector restructuring are a key but are not sufficient in themselves for the improvement of energy intensities' (p. 2295). While emphasising the central importance of energy efficiency for the development of more sustainable energy practices across the entire post-socialist realm, Ürge-Vorsatz et al. argue that the early 1990s provided the ECE states with a unique window of opportunity for 'leapfrogging' their economies towards a more sustainable energy sector, due to some of the unique advantages granted by socialist-era legacies: extensive use of public transport, increased reliance on rail freight, the presence of DH and CHP systems and so on. However, most Central European countries have failed to capitalize on this opportunity, so that 'the energy efficiency gap between the EU-15 and CEE ... has not improved significantly in Central European countries' (ibid.). The authors argue that such possibilities still exist, however:

> in the slowly transforming economies such as Russia and other former Soviet Republics. However, such leap-frogging requires not only inventive and dedicated policy-makers who dare not to copy "Western" policies but tailor new ones to local conditions; but also the "West", especially multilateral financial institutions agencies, to acknowledge and promote different, new pathways of development (Ürge-Vorsatz et al., 2006: 2295).

Nevertheless, the authors of this paper point out that substantial energy efficiency improvements would only be felt over the medium and long term, even if the most radical policy reforms are implemented in the near future. By emphasising the importance of behavioural and organizational change, their work has helped foster an increased awareness of the embeddedness of energy reform in the broader institutional and spatial context, and the need for giving greater weight to social welfare and energy efficiency arguments.

Energy tariffs, welfare, and efficiency: new policy agendas

Even though post-socialist countries needed not 'experience large increases in inequality and poverty, given the political will to create adequate transfer systems and safety nets and the wealth to afford them' (Niggle, 1997: 337), most governments were unable to create adequate social welfare programmes for the compensation of disadvantaged households, in response to the rapid upward rebalancing of energy tariffs. Instead, they resorted to maintaining indirect subsidies by either preserving the old tariff structure or tolerating non-payment for energy services. This confirmed Kramer's (1991) prediction that governing elites would be 'reluctant to compromise any further their already precarious political legitimacy by enacting measures that, although they might promote the long-run welfare of the national economy, would redound to the short-run detriment of consumer welfare' (p. 74).

The pervasiveness of arrears and debts across the transition space is evidenced by fact that electricity bill collection rates in the poorer transformation economies rarely exceeded 60 per cent of payables during the 1990s (EBRD, 2000a). By being so widespread and deeply embedded within the economy, issues of energy non-payment and political resistance to tariff reform began to threaten the financial viability of the energy utilities themselves. The existence of this connection may have been one of the reasons why social arguments gradually began to be incorporated into mainstream discourses on energy reform. This trend was part of a broader change in the treatment of social exclusion within economic transformation policies, whereby social equity was 'no longer considered to be an obsolete demand that will harm efficiency' (Ferge, 1998: 33).

Thus, the EBRD's 2000 *Transition Report* devoted significant attention to the social welfare implications of energy tariff rebalancing, by pointing out that the 'relatively high tariff increase for residential consumers' could be ameliorated by 'maintaining programmes of targeted income support' to low income households. At the same time 'tariff increases, together with appropriate and effective support for low income groups, are necessary to provide the correct incentives for investment in loss reduction and to allow the development of a competitive market' (EBRD 2000, p. 10). The energy affordability issue received even greater attention in the 2001 *Transition Report*, whose overarching theme was 'energy in transition' (EBRD, 2001). The report devoted an entire chapter to end-use energy efficiency in post-socialism, including a discussion of the social welfare and economic effects of energy price increases. This was one the first occasions where a major international financial institution like the EBRD had treated poverty and energy efficiency in post-socialism side by side, in a single framework.

The inter-dependence of energy and poverty also received special attention in two World Bank-sponsored publications on social exclusion in transformation. A report titled *Making Transition Work for Everyone: Poverty and Inequality in Europe and Central Asia* (authored by Jones and Revenga, 2000) synthesized the findings of national-level poverty assessments in the region. It combined household survey data with 'extensive' qualitative studies, with the aim of analysing the 'nature and evolution of poverty and inequality in the region'. Having concluded that 'the increase in poverty and inequality ... over the past decade is as striking as it is unprecedented' the authors formulated a number of policy steps for reducing poverty and creating 'inclusive societies'. They proposed a two-tier approach, which included 'fostering institutions at the community, local, and national level that are accountable to and inclusive of all parts of society' alongside economic growth. Overall, the report demonstrated a pronounced institutionalist bias.

This document was followed by Lovei et al.'s (2000) research on *Maintaining Utility Services for the Poor: Policies and Practices in Central and Eastern Europe and the Former Soviet Union*, which emphasized that 'maintaining, rather than expanding access to utility services, is the main challenge' facing post-socialist governments. The authors developed 'a conceptual framework for the evaluation of utility subsidy mechanisms', which was used to assess 'the mechanisms currently in use in post-socialist countries'. Similar to Freund and Wallich (1996) and Dodonov et al. (2001), they critiqued across-the-board price subsidies and no-disconnection policies, for supporting 'the middle class rather than the poor' while exerting a detrimental impact on the financial health of energy utilities, industrial competitiveness, as well as local and central government budgets. The report also argued that governments should develop systematic approaches to select the most adequate subsidy mechanisms for protecting vulnerable households from energy price increases (Lovei et al., 2000, p. 15).

One of the weaker aspects of both studies, however, was their failure to examine the multiple 'market barriers that prevent consumers from taking advantage of cost-effective energy efficiency opportunities' (Strickland and Sturm, 1998: 877). The marginalization of energy efficiency issues in post-socialist energy policies was highlighted by Müller and Ott (1998) in the introdiuction to an in-depth review of the institutional dimensions of energy and environment issues across the post-socialist space. They argued that the lack of policy co-ordination in energy and environment sectors 'resulted in further divides in the areas of economic policy, privatization schemes, regulatory standards, investment, technological transfers, property rights, security and public participation' (p. 11). This echoed the conclusions of the broader literature on energy efficiency (see for instance Martinot, 1998; Kazakevicius et al., 1998; Levine et al., 1991).

The mainstream theory of housing reforms in transition also paid increasing attention to issues of social equality. Experts in this field noted that 'restructuring the property rights of tenants has a number of important implications for the scale of subsidy and social equity' (Barlow, 1993: 57). It is worth noting the work of the Berlin-based European Academy for the Built Environment, which developed an extensive programme of study on, among other issues, the environmental and energy efficiency aspects of prefabricated panel estates (see EABE, 1996). At the same time, Clapham and Kintrea's (1996) research connected the post-socialist housing transformation process with decreasing social welfare and the removal of universal price subsidies:

> A key question for the future is whether low-income owners will be willing or able to keep their houses in good repair. One of the main beliefs behind the reforms is that low-income owners will be able to look after their own housing better than the state. But even assuming that ownership gives households a greater incentive to improve their housing and keep it in good repair, they need financial resources to do this (Clapham and Kintrea, 1996: 186).

Clearly, the late 1990s brought about a new awareness of the complexities of energy reforms in transition, and the need for sequencing restructuring efforts in the energy, social, environmental and housing spheres. However, as indicated by the two World Bank-sponsored studies cited above, disconnections still existed between the energy and poverty theories and policies, on the one hand, and housing transformation processes, on the other. This resulted in the relative neglect of efficiency issues

stands out for its comprehensive take on the relationship between energy, poverty and environmental problems. It is one of the few such studies to conceptualize the affordability of energy at the household scale within a services-based framework, having distinguished between indicators of the provision of energy services – such as fuel consumption and the utilization rates of energy devices available to the household – and measures of the sufficiency of energy services, including space heating, ventilation, domestic hot water and cooking (Kovačević, 2004). The study is based on a wide range of information about the energy consumption patterns of the population, combined with evidence about environmental and economic trends – a refreshing move away from the economistic readings of energy affordability that have dominated most studies in the field to date. One of its key conclusions is that:

> Energy use can be improved in Serbia and Montenegro by severing the nexus between the government and the energy companies that have captured it, unbundling services performed by vertically and horizontally integrated energy utilities and lowering barriers to entry. The main new entrants should come from the civil sector: consumer associations, professional organizations and environmental and public interest groups. People have to be concerned about their living conditions and empowered and encouraged to participate in development processes (Kovačević, 2004, p. 95).

To a certain extent, the methodological approach of this research is mirrored by another UNDP-financed study of energy affordability, this time in the Mongolian capital Ulan-Bataar (CSD, 2005). Based on a survey of 300 urban households, the authors have found that more than 40 per cent of interviewees weren't able to meet their basic needs, including energy. They underline the need for co-ordinating 'electricity prices and tariffs with social welfare policies and sector technical reforms, and the harmonization between utility consumption and policies on the reduction of air and environment pollution' (p. 67) This is unlike Molnar's (2002) work on 'the social and environmental impacts of power sector reform in Hungary', which mainly stresses the economic issues related to energy reforms. He points out that 'the poorest 10–20, perhaps 30 per cent of the population is impacted by energy price increases' (p. 6).

A report on *Power Sector Affordability in South East Europe*, subcontracted for the EBRD by a private consultancy – IPA Energy – has been the first regional study of this kind to incorporate an explicit conceptualization of energy poverty in its methodological framework. Having undertaken a series of analyses of common indicators of electricity tariff affordability, poverty and welfare, the report 'confirmed that power affordability was a problem for many consumer groups in South East Europe (for example pensioners, the unemployed, low income households). At the same time many of the South East European countries have not yet developed adequate social safety mechanisms to protect energy poor consumers' (EBRD, 2003, p. 2). Here, too, however, there is a paucity of integrated spatial-institutional understandings of energy poverty, as issues of energy efficiency and organizational change continue to be marginalized.

A further piece of EBRD-related energy affordability research has been published by Fankhauser and Tepic (2005), who estimated the vulnerability of low-income households to energy price increases by examining how energy burdens would change across 27 post-socialist countries in ECE and FSU under a hypothetical scenario where 'all utility prices are raised steadily to reach full cost recovery levels by 2007' (p. 15). Their study does not cover energy consumption as a whole, but rather concentrates on electricity and DH, in addition to water affordability. Besides pointing to the 'urgent need to improve social safety provisions more than they imply a need to postpone tariff reform' (p. 26), this publication is noteworthy for being one of the few – if not only – of its kind to use the term 'energy poverty' in the post-socialist context. Of no less importance is the suggestion that targeted efficiency programmes may be used as a poverty alleviation method, alongside the conclusion that future schemes to improve energy affordability will have to rely on 'a deeper understanding of affordability constraints and of energy and water poverty more generally' (p. 26). Having noted that 'it is surprising how little we still know about the consumption patterns and well-being of low income households' (ibid.), the authors point to the need for further conceptual and empirical work for designing or refining social safety mechanisms.

In their entirety, these new lines of research indicate that energy poverty is beginning to be recognized as a major issue within mainstream thinking about policy and theory in post-socialism. The times when energy affordability and efficiency in the domestic domain were treated as residual problems, to be treated by 'another' policy domain, are over. They have given sway to a more nuanced and sophisticated understanding of post-socialist transformation, one that recognizes that 'networked' problems at the borders of several policy domains require comprehensive action across government departments, private sector organizations and civil society. However, it remains unclear how and to what extent this increased awareness of energy poverty at the theoretical level will translate into an effective policy praxis. And, as I point out in the following section, even the latest body of research still has to address some key issues.

Where next? Future research and policy challenges

This chapter has looked at the manner in which theoretical and policy views on energy, social welfare and housing reforms in the post-socialist transformation have changed over the last 15 years. A review of the key theoretical literatures on economic restructuring revealed that the broader tensions between neoliberal and 'gradualist' theories about the transition were also felt within the energy policy domain. This is because the neoliberal model of economic reform, which was adopted by most governments in their efforts to transform the energy sector, failed to take on board more gradualist and 'institutionalist' arguments about the need to implement fundamental changes in the structure of the economy and society. This would also include the careful co-ordination and sequencing of energy, social and housing reform policies. In addition to aiding the expansion of hardship and poverty, the failure to develop adequate social safety nets for energy price increases also led to the emergence of non-payment practices and political resistance to further energy reforms. However, the last few years have seen a new generation of academic and policy work aimed at alleviating these policy traps.

Yet a number of questions and challenges remain open. For example, as pointed out in the previous chapter, one of the key issues is the need for developing a comprehensive, customized conceptualization of the term 'energy poverty' in the post-socialist context. Although it has been established that energy poverty-causing conditions, such as low residential energy efficiency, non-payment and income inequality, exist side-by-side in many Eastern European countries, very few empirical studies in this domain operate with the concept of 'energy poverty'. Instead, less comprehensive terms are being employed, ranging from 'energy disconnection' or 'energy non-payment', to 'poor living conditions' and 'social exclusion' (see Dodonov et al., 2001; Freund and Wallich, 1997). The failure to perceive all energy affordability-related problems under the common heading of 'energy poverty' may prevent scientists and policy makers from seeing its contingencies and implications in an integrated manner. Reducing the scope of analysis onto a limited set of symptoms (such as 'non-payment' or the 'bad living conditions' of the energy poor) obscures the broader spatial and political implications of the problem.

Moreover, many studies fail to make a connection between household-level processes, and broader political, cultural and economic trends. Although the late 1990s have seen an increased awareness of the environmental and economic dimensions of inefficient energy use at the household scale, much remains to be done in terms of incorporating this argument into the mainstream discourse on housing and social welfare transformation. In particular, the post-socialist poverty literature remains oblivious to the spatial and political factors that drive energy poverty, as social exclusion continues to be seen mostly as an income- and expenditure-driven problem. This is not to say that the spatial and institutional dimensions of poverty have been ignored by transition experts, as, for example, the housing literature regularly devotes attention to the problem of capital starvation in the residential infrastructures of poor households and communities (Pichler-Milanović, 1994). A vein of work on the 'geographies of the everyday life' in the post-socialist context (Stenning, 2005) open promising opportunities in this direction. Such insights have the potential to foster deeper understanding of the spatial-institutional production of social exclusion in post-socialism, if they are integrated within the conceptual realm of 'energy poverty'.

Last but not least, it has transpired that even the latest generation of post-socialist poverty research assesses social protection mechanisms via a reductionist view of poverty. Freund and Wallich (1996), Lovei et al. (2000), Jones and Revenga (2000) and World Bank (2003) all divide the population into 'poor' and 'non-poor' when assessing the social welfare implications of energy pricing. This 'black and white' separation is based solely on average household income criteria, in either aggregate or equivalent terms. The grouping of households into two discrete income categories is problematic, to say the least, because there is no consensus about when a household stops being 'poor' and becomes 'non-poor'. This sheds some doubt over the conclusion in Lovei et al. (2000), that the main consumer energy subsidy mechanisms used in post-socialist countries are primarily targeted toward the 'middle class', which they classify as 'non-poor'. A different poverty assessment method may have placed many of the middle–income strata below the poverty line.

However, a more fine-grained analysis of the multiple relationships between society, economy and space within the context of energy poverty requires an in-depth look at the manner in which household-level experiences of domestic energy deprivation are situated within broader dynamics of institutional, economic and political change. The remainder of the book is devoted to a study of these processes in Macedonia and the Czech Republic. Before delving into smaller scales, however, I will first provide a broad-scale assessment of the current state of knowledge about the patterns and structures of energy poverty across the post-socialist realm, in order to provide the background framework for the country-level analyses that follow.

Chapter 3

Patterns of Domestic Energy Deprivation Across the Post-Socialist Space

This chapter investigates the socio-spatial variation of energy poverty factors in ECE and FSU. It aims to explore, within the constraints of data availability, the manner in which the forces that shape the spatial distribution of inadequately heated homes vary across different spatial scales and social groups. Creating such an all-encompassing overview, however, has been difficult due to the absence of energy poverty from the conceptual language of both official data-gathering organizations and other research bodies in the region. With the exception of the handful of studies described in the previous chapter, there has been no work aimed specifically at establishing exactly which populations and locales might be affected by domestic energy deprivation. Aside from the poor political awareness about the problem, one of the reasons for this situation is the complete lack of data about domestic energy deprivation.

I have thus decided to construct a broad regional survey of the spaces of energy poverty by assembling various strands of indirect evidence from secondary energy, housing, health, and social welfare literatures, in addition to making a few cross-comparisons of my own based on data from several international organizations. But the gaping holes in some of the datasets, and the mutual incomparability of many of the sources that I have cited, lead to a sketchy and fragmented image. Hence, the aim of the chapter is not – and cannot be – a comprehensive overview of the 'geographies' of energy poverty in ECE and FSU. Rather, it aims to highlight how the generic factors and dynamics that shape the spatial distribution of inadequately heated homes vary across society and space. In this, it should provide the groundwork for the two subsequent in-depth case studies, by allowing further micro-scale investigations to be 'nested' within a broader context.

The chapter looks at the spatial variation of a selected group of trends that, according to the literature (for example, Boardman, 1991), are deemed to affect the extent of energy poverty. They include: general income poverty, energy sector reform, energy efficiency, non-payment, household energy expenditure and social assistance for poor households. It has been found that such factors vary dramatically across not only countries and regions, but also the towns and the neighbourhoods within them. This means that energy poverty affects different populations in widely divergent ways, and that there is a need for further in-depth investigations of its micro-scale patterns and experiences.

Indeed, Chapters 4, 5 and 6 are aimed at undertaking such investigations. However, in order to create the groundwork for more fine-grained analyses, one of the final outputs of this chapter is a grouping of energy poverty patterns into three generic types, based on the macro-scale distribution of the factors that have been surveyed: 'pervasive' (where domestic energy deprivation is a widespread problem in society), 'insular' (domestic energy deprivation is concentrated only among particular demographic groups) and 'potential' (domestic energy deprivation may develop in the future, after the removal of price subsidies). Such categories are closely linked to the structural outcomes of energy sector reform, which in turn reflect the macroeconomic and institutional conditions of post-socialist restructuring. This is consistent with the EBRD's assessment of 'transition progress' in various countries, which has concluded that 'The correlation between general progress in transformation and progress in energy reform indicates that the most successful reform in the power sector has been achieved against a background of progress in general reform, which has fostered economic stability' (EBRD, 1999, p. 7).

In the chapter, 'energy sector reform' is understood to comprise the restructuring of energy consumption-related operations, including electricity, district heating, and gas, although there is a heightened focus on electricity and district heating, because of their dominant share in the residential fuel mix.

Poverty in transition: broad-level structures and trends

The social composition and spatial distribution of general income poverty and inequality may contribute to the variation of energy poverty at different scales, because the inability to afford adequate warmth is related to broader patterns of social exclusion and deprivation (based on, for example, arguments in EBRD, 2003). However, a more comprehensive discussion of marginality in post-socialism is beyond the remit of this book. In the following sections, I chart some of the key underlying features of social inequality and poverty in transition, while outlining the multiple ways in which they could have contributed to the rise of energy poverty.

In order to understand the underpinnings of deprivation in post-socialism it is necessary to look back at the kinds of dynamics that shaped poverty during the socialist period. Permeating the literature on income inequality and social exclusion prior to 1990 is the conclusion that 'the national economies in the Central European countries had stalled long before the transitions began in 1989' (Torrey et al., 1999). For instance, Milanović's research of poverty in Yugoslavia, Poland and Hungary indicated that 'during the 1970s, poverty was a rural phenomenon in Poland, Hungary and Yugoslavia, but by the end of the 1980s it had become an urban phenomenon' (Milanović, 1991: 197). However, he points out that, unlike the previous decades, increased urban poverty in the 1980s did not occur as a result of rapid migration to the cities. Rather, it was caused by the gradual impoverishment of existing urban labour, reflected the inability to procure replacements for worn out consumer durables: 'TVs and washing machines became too expensive, and rising electricity bills and rents compressed the affordable standard of living below the accustomed level' (ibid.).

Milanović found that the poverty rate for the non-agricultural population (workers and pensioners) in the former Yugoslavia doubled between 1978 and 1987, although it remained lower than that of mixed households and farmers. The poverty rate of farmers declined from 42 per cent in 1978 to 27 per cent in 1983, but increased to 45 per cent in 1986 and 1987 due to changes in state subsidies (ibid.). But this may not be a universal pattern, as 'survey data suggest that the relative individual income transitions in Central Europe were less dramatic than the macroeconomic transitions of the individual economies, and the individual income transitions ... differed from country to country in terms of which groups suffered' (Torrey et al., 1999: 252). Nevertheless, it has been established that post-socialist states experienced a sharp increase in unemployment during the late 1980s, exceeding for Hungary and Poland the average rate of OECD countries (Atkinson and Micklewright, 1992; Eatwell et al., 2000). As a whole, these findings indicate that social stratification also existed under Communist rule, which means that ECE and FSU states inherited a pattern of income inequality in the late 1980s.

Poverty-inducing dynamics in post-socialism

When the transformation process began in 1990, it was expected that economic and political restructuring would take a short time, and that the benefits of GDP growth would soon 'trickle down' to the majority of the population. However, the experience of the last 15 years points to a more mixed picture: 'while some [transition economies] have survived the transformation in good shape, for others the consequences have been devastating' (Redmond and Hutton, 2000, p. 2). Even the most developed post-socialist states, however, have not been immune to the rise of income inequality: according to Adam (1996), the social costs of the transformation in Poland, the Czech Republic and Hungary 'were quite high and were not distributed equally' (p. 7). The poverty burden was 'borne by the masses', because 'the neo-liberal strategy was carried out in its pure form in Poland and in a moderate form in the Czechoslovak Republic' (ibid.). He argues that the same happened in the case of Hungary, although its restructuring process was carried out in a more gradual manner (ibid.).

The rise of income poverty and social inequality in ECE and FSU has mainly been fuelled by the decline of the secondary sector, as a result of the closure of large loss-making industrial enterprises. Hence, 'the initial hopes ... for rapid transformation and economic prosperity have quickly been tempered by an unprecedented decline in output, employment and incomes, by the further worsening of certain social trends which had already emerged under socialist rule, a massive impoverishment and a severe health crisis' (Cornia, 1994: 579).

After the sharp increase in the early 1990s, unemployment and poverty have more or less remained unchanged in all of the less-advantaged Eastern European countries, where a stagnant pool of jobless and poor people has emerged (Bhayla and Lapeyre, 1999: 109). This means that 'while different patterns are emerging in the most recent dynamics of unemployment, the spread of long-term unemployment is common to all countries in that region' (Večerník, 1995, p. 15). Massive job losses have thus been the greatest shock in the transition towards a market economy, and are a key characteristic of the evolution of the labour market (UNDP, 2003; Zoon, 2001; Götting, 1994). They have also driven the inequality of wage distribution in post-socialism. According to Milanović, the non-wage private sector contributed strongly to inequality only in Latvia and Russia. Pensions, paradoxically, also pushed inequality up in Central Europe, while non-pension social transfers were too small everywhere and too poorly focused to make much difference (Milanović, 1999). In this context, it is worth noting Jones and Revenga's (2000) conclusion that one of the main features of the social groups suffering from low energy affordability is the 'loss of stable income', due to arrears in wages, or labour migration from the secondary sector towards self-employment in agriculture and services.

However, the reasons for the emergence of poverty should also be sought on the consumption side. According to Kuddo (1995) '… price shock, in connection with many other economic and social aspects of the transition, has meant real tragedy for many families' (p. 65). Cornia (1994) has concluded that 'only in the case of Czechoslovakia did the initial wage increases granted to offset the expected price explosion prove adequate' (p. 579). The removal of socialist-era consumer subsidies and the lifting of price controls has played an important poverty-inducing role in this process. In particular, vulnerable households have suffered from the dismantling of removal of state support for housing, energy and transport services.

Identifying vulnerable groups: towards a regional profile

Although it is widely accepted that 'the transition to a market economy would benefit some groups more than others' (Atkinson and Micklewright, 1992, p. 1), there is considerable disagreement as to whether a common demographic poverty profile exists across all post-socialist states. Nonetheless, some key trends can be extrapolated:

> Hardest hit are large families; the disabled; the unemployed; pensioners; families of low paid workers; and individuals who have no access to land or other alternative sources of incomes. The entire population has been forced to completely change their way of life and consumption patterns, adapt to new economic conditions, and to lower their living standards. The social class which has benefited in terms of income from the transition constitutes less than 10 per cent of the population (Kuddo, 1995: 65).

The rise of deprivation and poverty among certain social strata is linked to the fact that 'Vulnerable populations during severe economic changes tend to be those who have fewer labour-market skills or less mobility, such as children, the old, women, and single parents' (Torrey et al., 1999: 246). According to Cornia et al. (1996) 'While the economic difficulties … have also have affected the 'old poor' (members of large and single-parent families, people with disabilities, some minority groups and the elderly subsisting on minimum pensions), the biggest increases have been recorded among the 'new poor' (p. 164). This group usually includes young people in search of new employment, uncompensated or long-term unemployed, retrenched low-skilled workers, socially-excluded migrants and refugees, farmers (with the exception of under-developed agricultural economies in the Caucasus, Albania and Romania), as well as the ever-increasing numbers of 'working poor' and their dependants (ibid.). Much of the literature stresses that the 'newly-unemployed' are one of the most vulnerable groups, as post-socialist unemployment tends to be associated with lower-than-average levels of education and qualification, and higher-than-average age. Also, women are more likely to be unemployed in post-Communist Eastern Europe, and in many ways 'are bearing the brunt of economic restructuring' (Makkai, 1994).

A further disadvantaged stratum is composed of families with children, especially single parents. It is difficult to find a broad-level analysis of social deprivation which does not emphasize the difficult conditions of such families, especially among ethnic minority populations. Even in Russia, which has a specific poverty profile, it has been found that multiple-children and single-parent households are, respectively, two and three times more likely to be income poor than nuclear families (Kumssa and Jones, 1999). The situation of pensioners, however, remains unclear. While some authors argue that 'pensions appear to have been better protected from inflation than other transfers' (Cornia et al., 1996: 164), so that, for instance, 'less than 1 per cent of pensioners in Russia are persistently poor' (Lokshin and Popkin, 1999), it should also be pointed out that 'elderly people depending on extremely low minimum pensions as their sole source of income have fallen into extreme poverty' (Cornia et al., 1996: 164). Such a situation has been found to exist in Russia, Ukraine, the Czech Republic and Poland (Plotnikova, 2002; UNICEF, 2000; Barr, 1996).

Conversely, there is little disagreement over the fact that working-age adults are in the best relative position *vis-à-vis* income inequality and wage distribution. This holds true in both the ECE and FSU context, where 'prime-age adults are likely to be in the always-rich category' (Lokshin and Popkin, 1999). Moreover: 'the highest deciles have lost less than the lowest deciles, and as time goes by they are even gaining: a process that will probably gain momentum as economic growth continues' (Caselli and Battini, 1997: 15)

Although they have provided useful insights into the systemic features of post-socialist poverty, these generalizations should nonetheless be interpreted with caution, because the extent and structure of poverty in the lesser-developed transformation states remains largely unknown. There are some indications that the patterns of social exclusion and deprivation in countries like Russia and Ukraine do not conform to those observed further west in ECE (Eatwell et al., 2000; Milanović, 1999; Ravallion and Lokshin, 1999). For example, it has been found that approximately 5 per cent of the Russian population is at risk of falling into extreme poverty: these are the people who 'have not fallen but are on the verge, the most vulnerable group within the social framework'. Moreover, 'one third of the people in this stratum have higher university education or even academic status. Their social relations have broken off, and morale is low, characterized by a high degree of fear and despair, they feel permanent stress and anxiety; uneasy about wage arrears rather than low rates of payment' (Rimashevskaya and Voitenkova, 2000: 49). On the other hand:

> The bottom stratum of the population in Russia – the socially excluded – is extensive. At present it amounts to 14-15 million people, about 10 per cent of the total population and compares with about 5 per cent in the United States. The continuing impoverishment and marginalization of significant sections of the population means that the numbers in the lowest stratum continue to grow, especially the numbers of children, for whom the future is uncertain' (Rimashevskaya and Voitenkova, 2000:49)

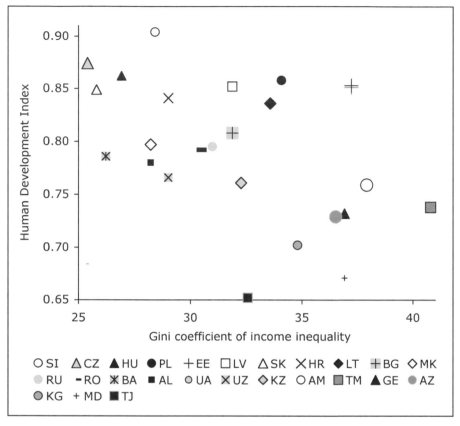

Figure 3.1   Relationship between the human development index and the Gini coefficient of income equality

Source: UNDP, 2005.
Note: AL = Albania; AM = Armenia; AZ = Azerbaijan; BA = Bosnia and Herzegovina; BG = Bulgaria; BY = Belarus; CZ = Czech Republic; EE = Estonia; GE = Georgia; HR = Croatia; HU = Hungary; KZ = Kazakhstan; LT = Lithuania; LV = Latvia; MD = Moldova; MK = Macedonia; PL = Poland; RO = Romania; RU = Russia; SI = Slovenia; SK = Slovakia; TJ = Tajikistan; TM = Turkmenistan; UA = Ukraine; UZ = Uzbekistan.

Macro-scale patterns of human development and income inequality

Different post-socialist countries have dealt with the transition crisis in very different ways, while inheriting different levels of development. It can thus be expected that rates of poverty and inequality would vary significantly between them.

   This is confirmed by the ranking of ECE and FSU states according to the UN Human Development Index (HDI). This is a composite measure of poverty, literacy, education, life expectancy, childbirth and the standard of living worldwide (UNDP, 2005). Despite its reductionist character, it provides interesting insights into the kinds of trans-national regional groupings that may have emerged in the post-socialist period. The most developed country – Slovenia – has an HDI score of 0.904, which gives it a global HDI rank of 26, while Tajikistan is at the bottom with an HDI of 0.652 and a global ranking of 122 (overall, the highest ranking state is Norway where the HDI amounts to 0.963, while the lowest HDI, 0.281, is Niger's at the 177th place). It is interesting to note that the EU accession states, with the exception of Romania and the addition of Croatia, have been ranked in the group of countries with 'high' levels of human development, as all of them have HDIs above 0.8.

Plotting the HDI against the Gini coefficient of income inequality – which measures the extent to which the distribution of income (or consumption) among individuals or households within a country deviates from a perfectly equal distribution – reveals two distinct clusters of post-socialist countries. The first of these includes all the Caucasus states, three of the Central Asian FSU republics (Kyrgysztan, Turkmenistan, Tajikistan) as well as Moldova and Albania (Figure 3.1). According to the definition given by the UN (see UNDP, 2005) such countries have low levels of 'human development' and high degrees of income inequality. The second cluster includes all remaining states, with a distinct grouping of Slovenia, Hungary, the Czech Republic and Slovakia at the top. All of these states have managed to achieve a relatively high HDI score without significantly compromising the equality of income. Estonia is a slight exception from this group, as its Gini coefficient resembles that of the lesser-developed Central Asian republics.

However, such broad national-level trends hide the more intricate spatial patterns of human insecurity and poverty that exist at the regional, urban and household scales. While a more detailed investigation of these patterns is beyond the remit of this book, it is worth noting that there are increasing numbers of the 'new poor' in the larger cities of 'lagging' transformation states. This group tends to consist of the general categories of vulnerable households listed in the previous section, as well as recent rural-to-urban migrants with limited opportunities on the job market (Struyk et al., 1996). Such trends have transpired despite the fact that capital cities in transition states have usually had the lowest poverty rates.

Thus, the judgement is still out as to how macro-scale social developments are reflected at the level of urban systems. While there has been a significant literature on the new trends of regional differentiation, urban social inequality and social segregation in transition, it remains unclear whether and how these developments can be generalized to fit the broader profile. Much of the literature stresses that pre-1990 patterns of socio-spatial segregation are being decomposed at different rates. While large cities are moving towards clear spatial structures of social inequality, it appears that lower-ranking cities have not yet reached the phase of distinct 'seas' or 'pockets' of poverty, because social exclusion has retained the spatially-dispersed character of the pre-transformation period (EABE, 2000; Sykora et al., 2000; Hampl et al., 1999). The extent of poverty in rural areas also remains to be researched: although such areas in most countries have been historically the poorest, it appears that the 'primitivization' of some of the lesser-developed transition economies is reversing the traditional urban-rural poverty divide (Raiser et al., 2001; Potter and Unwin, 1995).

Geographies of energy transformation

As a whole, post-1990 energy reforms have created an additional geography of unevenness across ECE and FSU, because some countries have been more successful than others in adapting their legal frameworks to the new conditions created by the post-socialist transformation, and especially the decision to adopt the Anglo-Saxon mode of energy regulation. A complex web of conflicting forces and interests has directed

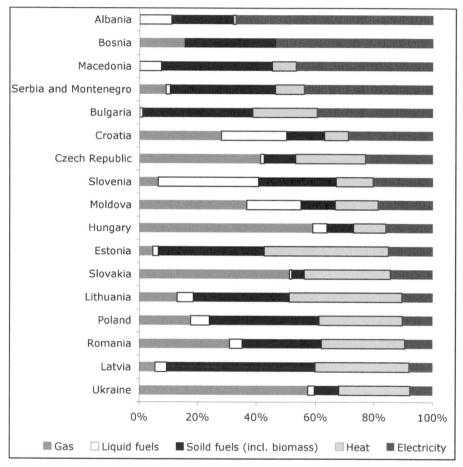

Figure 3.2   Structure of total final residential consumption by fuel

Source: AEA, 2006.

the restructuring process into divergent paths, while also influencing the manner in which energy poverty may be spread across space. Although the constraints of this book do not allow for a detailed, up-to-date review of energy reforms in each country, it is still possible to chart some of the common geographical underpinnings of energy sector reform in ECE and FSU, while noting the sub-regional groupings that may have emerged as a result of differences in the nature of economic and legal restructuring.

Even though, as pointed out in the introduction, the energy sectors of post-socialist countries encompass a variety of operations, including electricity, gas, DH (district heating) and oil, it is the transformation of the electricity sector that has accentuated the differences between countries most prominently. This is because old ownership and even regulatory structures are still widely present in DH, which continues to be municipally owned and managed throughout the post-socialist space. Despite the fact that most countries have moved their gas generation, transmission and distribution systems towards market-based models, even here the state continues to play a major role, as evidenced by the fusion of state and business in Russia's Gazprom, and to a lesser extent, Hungary's MOL.

As for the domain of oil operations, although it is now exposed, as it were, to the vagaries of the free market throughout the post-socialist space (see, for example, EBRD, 2001), the spatial ramifications of oil industry restructuring are too complex and extensive to be analysed within the constraints of this section, especially because they are not particularly relevant to household consumption where liquid fuels constitute only a minor share of total demand.

However, unlike all of these sectors, electricity reforms have followed dissimilar paths across different countries and regions. Given that this fuel dominates residential energy expenditure and consumption in most countries (see Figure 3.2) it is worth taking a look, at least, at the macro-level patterns of its systemic restructuring.

One of the shared features of nearly all electricity sectors in the region is the overwhelming entry of private – and mostly foreign – capital post-1990, mainly in the form of new independent power providers (that is, privately-owned generation facilities supplying power to the national grid), and purchases of low-voltage distribution companies (the so-called 'Discos'). There is hardly a Central European, Balkan or Baltic post-socialist state that has not sold some or all of its discos to foreign and/or domestic private companies. A number of Polish and Ukranian distribution companies that were included in this process have been ranked among the largest privatizations in the post-socialist transition (see Burdett, 2002).

However, privatization also requires the 'unbundling' and liberalization of energy operations, which has, in some cases, created financial and technical difficulties for state-run electricity transmission grids. The problem can be found in very different transformation countries, regardless of the phase of macroeconomic reform. Transmission companies owned by the state in Hungary, the Czech Republic, Ukraine, Moldova and Russia have all been reported to be facing poor solvency problems. One of the causal factors for this situation has been the emergence of a 'black market' for electricity, in the form of direct supply relationships between generators and final distributors, and negotiated access to the transmission lines (Commander et al., 2002; Opitz, 2000; Allsopp and Kierzkowski, 1997). The problem is further aggravated by the 'barterization' of post-Soviet economies, where semi-legal energy transactions have usually been compensated with non-monetary means (ibid., and Siner and Stern, 2001).

Apart from these two trends, however, it is difficult to speak of any other common characteristics of electricity reform in transition. Yet there are macro-regional differences and similarities between countries that will be explored here through the lens of broader energy transformations.

Extensive reforms in the EU accession states

It has been claimed that the implementation of neoliberal energy regulation in the transformation economies with stronger legal cultures has resulted in the evolution of an intermediate – and yet stable – mode of regulation: the 'Central European' model. This is because many Central European 'governments were able to implement coherent strategies, energy companies have been set up that operate according to established legal and economic criteria, and a certain separation between them and the state was institutionalized' (Wälde and von Hirschhausen, 1998, p. 14). Indeed, most Central European countries were among the first to unbundle their electricity sectors, while attracting substantial private investment from abroad (World Bank, 1999b; 2000). This is particularly true in the cases of Poland, Hungary and Estonia, whose electric power markets reached opening rates of more than 50 per cent by 2002 (PiEE, 1999; 2002).

Hungary has been a showcase of neoliberal reforms, because the liberalization and privatization of its energy sector commenced already in 1991 (EBRD, 2000b). Aside from international financial institutions like the World Bank and EBRD, this process has been mainly driven and motivated by the European Union, as evidenced by the fact that 'harmonization with and eventual accession to the EU' was one of the main objectives of the Hungarian energy policy, adopted in 1993 (IEA, 2000, p. 155). In 1995 and 1996, the Hungarian Government sold public shares in the electricity distribution companies, as well as four out of seven electricity production facilities (HEO, 2000). These shareholdings were sold to foreign investors – mostly public utilities in Western Europe – although the state maintained a 'golden share' in all the privatized companies, 'to allow the Government to veto decisions on capital increases, share issuing, changes of activity, mergers and de-mergers' (ibid.). Thus, most of the Hungarian electricity system was unbundled and privatized by the end of 2000. But the rapid liberalization process has been associated with political controversy:

> The Hungarian ministry of economic affairs has announced changes to the basis of energy prices which are due to take effect from 1 July [1999]. The new system [shifts] ... the balance of prices ... onto household and small users ... The current usage band system ... will be abolished ... Although the Hungarian Energy Office (HEO) states that changes will mean an average increase in end user prices of 2-3 per cent, it is clear the increase for users at the lower end of the band is 11 per cent – as the HEO itself admits ... This is surprising in view of the government's pre-election promise last May to protect the interests of the "ordinary Magyar". Clearly the big gainers in the new structure are users in the highest band, above 300 kWh/month, which does not include many ordinary Magyars living on a salary of $150 per month (PiEE, 1999).

Unlike Hungary, however, the governments of the Czech Republic and Slovakia have been less willing to relinquish central control of the energy sector: 'The Czech Republic and Slovakia fell behind their neighbours in energy privatization because of vested interests, nationalist concern over the industries' strategic importance, and the potential political impact of higher bills' (Anderson, 2001:15). The reluctance of in energy service provision to the poor. Moreover, social welfare arguments were poorly incorporated within the dominant discourses on energy and environment.

Re-thinking energy affordability

The recent years have seen a new generation of research and policy studies on the social effects of energy reforms in transition. This work displays a more refined understanding of the economic, geographic and historical contingencies of domestic energy deprivation, while operating with a comprehensive analytical approach to study it. One of the most notable examples of such thinking can be found in yet another World Bank-sponsored study, titled *Coping with the Cold: Heating Strategies for Eastern Europe and Central Asia's Urban Poor* (Lampietti and Meyer, 2003). It examines the relation between heating, poverty alleviation and environmental quality via an econometric analysis of the energy demand and prices. The authors use household survey data from Armenia, Croatia, the Kyrgyz Republic, Latvia, Lithuania, Moldova and Tajikistan, in order to analyse household energy consumption and heating patterns. The book concludes with a series of recommendations on how to design policies that will enable the provision of 'clean heat' in 'fiscally-sustainable ways' (ibid., p. 23).

Another elaborate investigation of the relationship between social welfare and energy sector reform in post-socialism has been provided by a United States Agency for International Development-financed study of the social support mechanisms for energy price rebalancing used by Armenia, Bulgaria, Hungary, Kazakhstan and Romania (Velody et al., 2003). It looks at how different countries have used three types of mechanisms – fuel assistance payments, energy efficiency improvements in low-income residences and 'progressive' tariff structures[1] – as means of alleviating the hardship faced by disadvantaged households in response to electricity price rebalancing. Having discovered that 'energy costs are the highest monthly expense after food for most low-income households in the region' (p. vii) the report concludes that the most successful energy assistance programmes were those which target vulnerable households directly, provide a meaningful level of protection, are not combined with other social welfare schemes and are simple and easy to administer. The authors also emphasize that:

> energy efficiency can and ought to be an important element in helping poor households when utility costs to consumers increase. Energy efficiency measures not only provide considerable cost savings to poor households, but also allow governments to maintain the adequacy of existing energy assistance payments to the poor, while strengthening incentives to save energy during periods of price liberalization (Velody et al., 2003, p. 48).

The recent years have also seen several national-level studies of domestic energy deprivation. An UNDP-sponsored investigation in Serbia and Montenegro work

---

1   Tariff systems that charge a lesser amount for low electricity usage or usage at certain times of the day.

these states to relinquish price-setting power to an independent regulatory body may have been prompted by the high social costs of energy tariff reforms.

Some of the EU accession countries further East – including Lithuania, Latvia, Bulgaria and Romania – have also been unwilling to embark on radical energy reforms. This has transpired despite strong outside pressure, especially from the EU (Balmaceda, 2002). Latvia experienced strong political opposition to utility unbundling and privatization:

> Latvenergo, the integrated state energy monopoly, was to be privatised by the middle of 1998, but a lack of political consensus over its restructuring and the privatisation method to be used contributed to long delays ... In February 2000, the government approved a plan for the restructuring of Latvenergo by the end of the year and the privatisation of some of its parts in the first quarter of 2001 ... However, 23 per cent of eligible Latvians signed a petition for a referendum against privatisation and in August 2000, in accordance with the results of the referendum, parliament adopted amendments to the energy law blocking the sale of any parts of Latvenergo, retaining it as a vertically integrated company (EBRD, 2000c, p. 23).

Bulgaria has undergone similar problems, as governments with a more radical reform approach have been regularly voted out of office. Although most energy restructuring steps were implemented by Ivan Kostov's government, which exposed the economy to a rigorous set of IMF-style policies in 1997 and 1998, the 2001 parliamentary elections brought about a record loss for his United Democratic Forces (Alexandrova, 2000: 14).

Nevertheless, as a whole, the 10 countries that have joined the EU have progressed the furthest in restructuring their energy sectors. Most of them have implemented the EU electricity liberalization directive in full, having also opened energy markets and changed the ownership and regulatory structure of formerly state-owned electricity monopolies. Whether this has been followed by adequate social policies, however, is less clear.

Sluggish reforms in the Balkans

The non-EU states of Southeastern Europe – the former Yugoslav republics, plus Albania and Moldova – form a distinct geographical grouping with respect to the path of energy restructuring. Even though reform patterns in the region vary between countries and sectors, all such states are distinguished by two common features: the poor institutional and regulatory capacity of the central state, and the overall low level of economic development. Most of them have retained state-owned and vertically-integrated energy utilities, despite efforts to pave the road towards full restructuring, commercialization, corporatization and privatization. Governments have been reluctant to relinquish central control over energy monopolies in order to retain political and strategic power over them (EBRD, 1999, Annex 9, p. 1).

The restructuring process has also been hampered by vested interests, over-employment in energy companies and indirect price subsidies. Problems of non-payment and electricity theft – especially in Albania as a consequence of the general poverty of the population and the financial indigence of large industrial enterprises –

are another inhibiting factor towards comprehensive energy reform (Polackova Brixi et al., 1999; World Bank, 2000). These difficulties have persisted besides the fact that many Balkan countries have already seen foreign private interest in the energy sector, involving companies such as America's AES, Entergy, Horizon Energy, the German RWE, Italy's Enel, as well as Japan's Mitsui, among others. International agencies – including the EBRD, World Bank and USAID – have been supporting this process through various small scale initiatives, feasibility studies and advice to governments (WIIW, 2004) In recent years, the Balkan states have moved towards the creation of a common power market, allowing electricity to be freely traded across national boundaries (AEA, 2006).

Despite its wide variety of resources, this region as a whole is a net energy importer. The lack of energy is particularly pronounced in the case of electricity, where, prior to 2006, most of the excess demand was met by Bulgaria. The EU-led decommissioning of reactors 3 and 4 of the Kozloduy Nuclear Power Plant in late 2006 led to severe shortages of electricity in several countries of the region, due to poor transmission infrastructures and high prices. Thus, during January 2007, Albania was only able to obtain about 70–80 per cent of its daily needs of 20 million kWh of electricity, despite importing about 10.7 million kWh per day. As a result, the country was forced to institute blackouts which lasted up to 15 hours in some rural areas. Even developed countries adjacent to the region, such as Greece, have begun to fear that they will be unable to provide electricity at times of peak demand as a result of the decreased supply of electricity in the region as a whole (Todorovska, 2007).

Collusion of state and business in the Former Soviet Union

Energy sector restructuring processes in Russia, Ukraine and the FSU republics in the Caucasus and Central Asia have generally followed a haphazard and chaotic path. The 'unbundling' of energy industries in such countries has effectively amounted to a superficial reorganization of operations, allowing 'real' decision-making power and profits to remain in the hands of governing elites (Kennedy, 2002a; 2002b; Stern and Davis, 1998; Braber and van Tongeren, 1996). At the same time, the privatization of energy companies in the countries with weak informal legal and political cultures has generally involved 'financial', rather than 'strategic' investment, invigorating shady business practices and mismanagement of the energy sector.

For instance, although Ukraine pledged to adopt a competitive pool regulation of the electricity sector in 1994 – nominally more liberalized than a similar scheme operating in the UK – subsequent half-hearted legislative reforms led to pervasive non-payment problems, barter chains, political interference and insolvency of energy enterprises (Dodonov et al., 2002). Although seven out of 27 Ukraine's *Oblenergos* (regional distribution companies) were already 'privatized' in 1998, they eventually came to be controlled by offshore companies with unclear capital bases and investment plans (Solomon and Foglesong, 2000). States in the Caucasus and Central Asia have been facing similar problems, as ruling elites have rushed to profit from the sale of strategic state assets in the energy sector, without recourse to the long-term consequences of such a move (Anex, 2002; Kreibig et al., 2001; Kaiser, 2000). For example, Kazakhstan's privatization of its thermal generation sector was described by a World Bank report as 'unplanned, rushed, opaque with little competition, and thus subject to allegations of corruption' (World Bank, 1999b, cited in Velody et al., 2003), while 'a handful of the many power distribution networks were sold to major international power companies, which began to implement substantial reforms, but have encountered regulatory problems including inadequate tariffs' (Velody *et al.*, 2003, p. 4).

At the same time, Russia's ability to transform its energy sector in a consistent and comprehensive manner has been hampered by the sheer size of its economy, the structural legacies of Communism, and the peculiarities of post-Soviet politics (see Rodionov, 1999). Such factors have combined to create a dual economic model for the electricity sector, in which elements of the Soviet system co-exist with modes of market and enterpreneurship (Opitz, 2000: 147). However, it remains unclear to what end this intermediate structure is merely a step towards a true neo-liberal market model or whether it is 'here to stay', in the form of a special Russian quasi-capitalist mode of energy sector regulation (for a wider discussion, see Bradshaw and Kirkow, 1998; Commander and Mumssen, 1998).

The complex tendencies and circumstances that have developed over the past 10 years may lead to a different categorization of the above countries, if other combinations of energy reform trends were to be considered. Nevertheless, the point that has been illustrated here is that progress in energy sector transformation generally mirrors regional restructuring trends, although there has been a time lag between economic reforms, on the one hand, and the implementation of a functioning legal and economic framework for energy operations, on the other. This gap is particularly pronounced in the former Soviet republics.

The efficiency of energy use: spatial and social differences

As was noted in the previous chapter, transition countries inherited a pattern of unsustainable energy use from the Communist system. The ability to move towards more efficient energy consumption practices at the household level can play a key role in preventing the emergence of energy poverty, because it decreases the demand for energy at the point of consumption, while improving the quality of the energy service. However, different states, regions and cities in ECE and FSU have responded differently to such policy challenges. This layer of spatial inequality is supplemented by a social one, which embodies the ability of various social groups to gain access to energy-saving technologies and infrastructures.

Despite numerous improvements of domestic energy efficiency, through, for example, the introduction of more efficient appliances or the refurbishment of building stocks and energy infrastructures, it is also true that unsustainable practices are still over-represented among low-income social strata. In per capita terms, 'poor' households have been found to consume, on average, at least 30 per cent more energy than 'non poor' households in Latvia and Lithuania (Lampietti and Meyer, 2002, p. 11). The same ratios exceed 25 per cent and 20 per cent in the cases of Moldova and Croatia, respectively (ibid.). In part, this can be explained by the continued persistence of socialist-era legacies, whereby more prosperous areas, and cities in general, had improved access to network energy such as DH or central gas, thus obtaining more efficient means of heating the home. Households not covered by these infrastructures tended to rely on less sustainable sources of warmth: 'the poor are more likely to use dirty fuels such as wood (Armenia) and coal (Moldova), while the nonpoor rely on clean fuels such as electricity and central gas' (Lampietti and Meyer, 2002, p. 13).

These trends have been reinforced by post-1990 developments, which have brought about a widespread movement towards gas-fired central heating systems in many urban areas throughout Central Europe. Wealthier households have switched away from DH towards individually-operated heating systems in order to benefit from the greater flexibility, reliability and cost-effectiveness that they provide (see Szomolányová et al., 1999; World Bank, 1998; Evans, 1995). DH use has also declined in Bulgaria, Moldova and Romania after increased disconnection rates due to non-payment, as well as Armenia and Georgia as a result of the political and economic crises in those countries. At the same time, disproportionately high numbers of households in some Balkan states have started to use electricity for domestic space heating, partly as a result of the under-pricing of electricity relative to other fuels (see World Bank, 2000). This kind of energy demand creates serious supply bottlenecks during periods of peak demand in winter. It is easily evident in Figure 3.2, where the top five countries according to the share of electricity in total residential energy demand are all in the Balkans. Conversely, gas household consumption dominates in Central Europe, while DH is the most widespread in the Baltics – partly as a a result of the Soviet legacy, as discussed in the previous chapter (also see Table 2.1).

- The spatial variation of energy efficiency practices and policies has multiple underpinnings that are too complex and extensive for the confines of this book. However, two of its facets have a particular significance for energy poverty, because they are directly related to the regulation of energy services at the household scale: The first aspect refers to ability of energy companies or consumers to measure the amount of energy used by households. The (non) introduction of energy meters in the residential sector shapes the spatial distribution of inadequately heated homes by influencing the manner in which households may or may not reduce their energy consumption, while acting is a key precondition for the overall marketization of the energy sector – since energy companies cannot function on an economic basis without proper energy metering.
- The second aspect of energy efficiency policies refers to the overall regulatory framework: energy poverty in a given locale is affected by the ability of governing agencies to provide sound institutional support for improved energy efficiency, which leads to demand reductions and improved service provision.

Geographical variation in the introduction of energy meters

In nearly all post-socialist countries, a key reform challenge towards the restructuring of energy operations at the demand-side of the sector has been posed by the lack of meters for the consumption of energy. As pointed out in the previous chapter, the socialist planners' understanding of energy as an unconditional right rather than a service associated with a certain price resulted in the omission of network gas, DH, and sometimes even electricity meters in many residential buildings. Lampietti and Meyer (2002) underline that 'virtually no heat or hot water metering existed before 1990 in residential, commercial and public sector buildings in most countries in Eastern Europe', as 'there was very little point in installing meters because consumers could not regulate the heat supply' (p. 51). Policy responses to address this problem have created a specific geography, as pointed out by Velody et al. (2003):

> At present, electricity metering is extremely widespread in the CEE region and Eurasia, but some meters are easy to tamper with so fraud or theft can be a problem ... Incidence of natural gas, district heating and water metering varies widely from country to country. Some buildings have no form of metering, others have one meter for the whole building, while others have individual household metering (p. 24).

Although there has been no comprehensive research on the current spatial incidence of different types of energy meters at various scales across the post-socialist space, in general it can be said that metering is more widespread in urban areas, as well as the 'advanced' post-socialist countries of CEE. This is because:

> Since 1990, many countries have made substantial progress in installing regulation and metering equipment. Many countries (Poland, Hungary, Bulgaria) have introduced mandatory metering at the building level, and in many cities in the Czech Republic, Hungary and Poland, for example, the metering rate is now close to 100 percent. In many countries of the FSU (excluding the Baltics), however, very few (under 1 percent) residential buildings have meters (Lampietti and Meyer, 2002, p. 51).

While it is true that the quicker pace of energy reforms has contributed to the recent improvement of metering capabilities in such areas – because it is impossible to establish a commercial, market-based framework for energy operations without knowing how much energy is consumed by each individual customer – this situation can also be explained by the longer history of metering practices in, especially, Central European cities, where the introduction of electricity and gas infrastructures before World War 2 was almost always followed by the installation of consumption measurement devices.

Thus, socialist-era housing estates in the FSU, as well as urban areas dependent on district heating, rank among the worst offenders in terms of the possibility of measuring household energy consumption. In locales like these, 'the legacy of the absence of metering has complicated reform efforts' (Velody et al., 2003, p. 25). It may have affected the extent of domestic energy deprivation, as low-income households 'do not bother to attempt to reduce energy consumption if the savings do not result in a clear financial benefit to the household' (ibid.). This is because it is impossible to quantify the reduction of energy consumption, and consequent reductions in energy expenditure, without being able to measure the exact amount of energy used by the household before and after the introduction of energy-saving technologies. For example, it has been found that low-income Armenian households 'with full control of their heating arrangements keep their apartments at lower temperatures and spend less than do households on the district heating network – suggesting that the poor lose the most when they cannot regulate their heating use' (Lampietti and Meyer, 2002, p. 15)

Energy efficiency institutions: a further layer of dissimilarity

The instalment of energy meters is only one facet of improved energy efficiency at the household scale, as a more comprehensive reduction of thermal losses in buildings requires wider programmes of window weatherization and improved wall insulation, among other steps. However, such measures require the development of adequate policy instruments by both the state and the private sector. A quick look at energy sector regulatory practices across the post-socialist space reveals that efficiency has received little policy importance in almost all post-Communist states. Only in recent years have decision-makers begun to see the broader benefits of moving towards a more sustainable energy economy. But even in the cases where there has been a general genuine political will to move towards greater energy efficiency in the household sector, reform attempts have often been stifled by a combination of institutional inertia and vested interests (Strickland and Sturm, 1998; Meessen, 1998). Velody et al.'s (2003) study of five post-socialist states concluded that:

> There is no evidence of governmental interest in promoting low-income energy efficiency in any of the countries studied. EU accession countries have adopted the EU appliance labeling system, and some have issued declarative energy-efficiency legislation, but with the notable exception of Hungary, there is little government financing for household energy-efficiency programs (p. 23).

To a certain extent, the link between energy efficiency and structural sector reform is exemplified by the relation between EBRD 'infrastructure transition indicators' and national-level electricity intensity during the first ten years of transition. The EBRD appraises the energy sector of each transformation country on an annual basis, assigning a score of 0 to 5 depending on a set of pre-defined criteria: '0' represents a 'perfect' non-reform situation (wherein socialist managerial and pricing structures have been retained in full), while '5' corresponds to the neoliberal Anglo-Saxon regulatory ideal (that is, the sector is fully liberalized, unbundled, independently regulated and controlled by private capital). It is indicative that the five countries whose infrastructure indicators reached 3 by 2001 all managed to reduce the electricity intensity of their economies during the first 10 years of transition (Figure 3.3).

However, beyond these five countries, it is difficult to find a relationship between infrastructure indicators and electricity intensity. To a certain extent, this may be attributed to the fact that decreases in the energy intensity per unit of GDP do not necessarily reflect effective progress towards an energy-efficient economy. This is

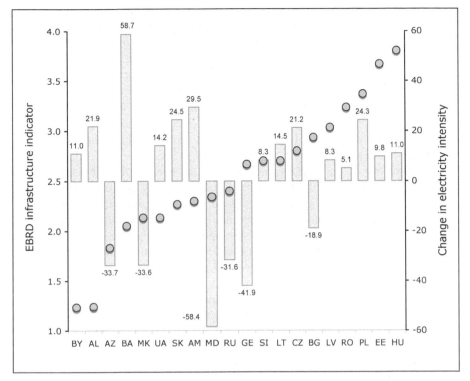

Figure 3.3 **Relationship between average EBRD infrastructure transition indicators during 1995–2001 and percentage changes in the electricity intensity of the economy between 1992–2001**

Sources: EBRD, 2001; OECD, 2003; 1994.

Note: The percentage change in electriciy intensity has been calculated by subtracting the 1992 figure for electricity intensity (defined as total electricity use per GDP) from the level in 2001, and dividing the difference by the 1992 figure. A negative percentage change implies greater energy intensity in 2001, i.e. a less energy-efficient economy.

immediately clear in the cases of Albania and Bosnia, where the apparent increase of energy efficiency was in fact caused by rapid economic contraction as a result of war and lawlessness, rather than structural changes. Although energy intensity is the best available proxy for systemic transformations in energy efficiency (see Lackó, 1997; PHARE Multi-Country Energy Programme, 1995) a more detailed look at the structure of the relevant policies at different scales is also needed.

Thus, in the 'advanced' transition states of Central Europe, energy efficiency improvements have often been driven by private investment. As pointed out by Kazakevicius et al., (1998) 'in both Hungary and Poland lighting market transformation towards more efficient technologies occurred under a strong market push from the manufacturer or international organizations. In both cases, government involvement was very limited' (p. 609). The same group of authors also emphasize that, in Lithuania, the 'lack of awareness as well as the initial high costs appear to be significant barriers to initial purchase' of compact fluorescent light-bulbs (CFLs). Furthermore 'state ownership and very limited financial autonomy of regional electric utilities restrict their abilities to participate in lighting efficiency programmes' (ibid.). Indeed, the broader literature on residential energy efficiency in transition indicates that governments have generally lacked the capacity to deal with the complex policy challenges of developing a comprehensive energy efficiency policy in the residential sector. One of the facilitating circumstances in their case has been the EU accession process, which has provided a key political and financial impetus for the establishment of comprehensive institutional structures (see Secrest, 2002; Evans, 2000).

In Poland, one of the crucial components of the energy efficiency policy was the Energy Law adopted in 2000 – a requirement for EU accession. Under this legislation, the Ministry of Economy assumed the main responsibility for energy policy, including energy efficiency. The country also strengthened the capacity of the main governmental body responsible for the operational implementation of energy efficiency policies: the Polish National Energy Conservation Agency, a joint stock company established in 1994. Other important actors included the Foundation for Energy Effective Utilization and various regional energy conservation agencies. The law also incorporated a requirement that local authorities prepare local energy development plans, to encourage the introduction of more sustainable practices and technologies at the local level (IEA, 2003).

End-use efficiency in the lesser-developed states of the Balkans and the Former Soviet Union has ranked even lower on the list of governmental policy priorities. A quick look at the main international surveys on energy efficiency reform in these contexts points to the state institutions and energy utilities' lack of willingness and capacity to develop and implement wider programmes of demand-side efficiency management; including, for instance, market transformation policies to replace the ageing stock of inefficient domestic appliances, or state-supported investment schemes for insulation improvements and weatherization (Lubinski, 1998; Klarer, 1997; Avdiushin, 1997). Multilateral development banks and bilateral agencies have played a key role in promoting domestic energy saving programmes in this region (see Strickland and Sturm, 1998; Guyett, 1991), partly because energy efficiency policies in the private sector have been constrained by the lack of investment capital. Governments have mainly directed budgetary funds towards generation or transmission infrastructure, where the most urgent repairs are needed (Wälde and von Hirschhausen, 1998; Gray, 1995).

The affordability of energy

> The energy bills of 90 per cent of Montenegrin households are higher than their salaries or pensions (Dnevnik, 2007).

The ability of households to pay for the energy services they receive is a crucial indicator of energy poverty, because it embodies the relationship between social welfare, energy prices, and household energy consumption. One of the key measures in this respect is the 'energy burden' – also known as the 'affordability ratio' or 'relative energy expenditure' – which is defined as the share of energy expenditure within total household expenditure or income. Looking at how energy burdens are distributed across space may provide important insights into the geographical distribution of domestic energy deprivation.

Yet here too there is a paucity of data. The only region-wide survey of post-socialist energy affordability to date has been Fankhauser and Tepic's (2005) EBRD-sponsored study, which calculated average energy burdens for 27 post-socialist countries in ECE and FSU from household expenditure survey data. Overall, their analysis indicated that 'Consumers in south eastern Europe (SEE), for example, spend on average more than twice as much on electric power as households in the Commonwealth of Independent States (CIS) and one-third more than households in central Europe and the Baltic states (CEB). At the same time SEE and CIS consumers spend considerably less on district heating' (p. 7).

The authors attribute these geographical differences to the fact that the restructuring of the electricity industry has progressed further than DH reforms, which means that power prices tend to much closer to cost recovery levels than DH, while collection rates in this sector are higher. At the same time, levels of electricity and heat consumption vary widely across the region. As noted above, electricity is often used for heating in locations where DH is under-developed or under-priced, such as the Balkans. Coupled with the relatively low incomes of the population, this is one of the main reasons why relative energy expenditures are so high in this region. At the same time higher tariffs 'are also partly responsible for the comparatively high heating bills in CEB' (p. 9). However, the authors are quick to point out that their analysis does not include expenditures on wood or coal, 'and therefore underestimates the total amount of money spent on heating' (ibid.).

Still, such national-level averages may hide significant differences not only within countries, but also across different social strata. The variation of energy burdens among different parts of the population has been explored by numerous authors, including, most notably, Fankhauser and Tepic (2005), Lampietti and Meyer (2002), the EBRD (2003), and Lux et al. (2003). The gist of this work is that low-income households tend to spend a higher share of their income on energy. For example, Lampietti and Meyer (2002) have found that the energy burdens of the 'poor' in Armenia were more than 30 per cent higher than those of the 'non poor', while the same ratio hovered around the 10 per cent and 20 per cent mark in Moldova and Lithuania, respectively. However, in Latvia, the poor used almost 50 per cent more of their income for energy expenditures than the 'non poor'. This is similar to Freund and Wallich's (1996) results for Poland and may reflect the higher energy burdens of pensioners, who, as was noted in the previous chapter, are over-represented among higher-income groups in the household expenditure survey.

In addition to the territorial distribution of energy burdens, the affordability of domestic warmth is also expressed in space through two other aspects of energy transformation: the emergence, in some countries, of debts and arrears towards electricity payments; and the different mechanisms used by different post-socialist countries in order to create a social safety net for energy price increases. The next two sections take a closer look at these issues.

Non-payment in the electricity sector

The gap between 'payables' (billed revenue) and 'receivables' (collected revenue) is one of the most tangible implications of energy poverty. In the electricity sector, it expresses the monetary power of households with respect to the energy services they receive, and the ability of energy companies to deal with debts and arrears in a prompt and effective manner. This issue has received considerable attention in the literature, because it is creating operational problems for energy utilities and governments. It too has a specific geography that may provide clues about the spatial concentration of domestic energy deprivation.

Although, in the main, non-payment embodies the poor affordability of energy services in post-socialism, it should also be pointed out that some of its causes are cultural. This is because, as pointed earlier, energy is seen as a universal entitlement, rather than a service to be paid for. Another reason could lie in the fact that 'the consumers find the energy service so poor, that they want to punish the companies by not paying' (Jørgensen, 2002, p. 15). However, poor collections have translated into a chain of inter-company arrears and debts, as well as wage arrears, while low levels of liquid capital in energy utilities lead to capital starvation and energy black-outs. It is paradoxical that many FSU states – particularly in the Caucasus – have been unable to supply their capital cities with more than two hours of electricity a day, despite surplus generating capacity, in addition to ample oil and gas reserves (Kaiser, 2000).

Another implication of non-payment is the increasing de-monetization of the economy (a pervasive problem in Russia and Ukraine), whereby economic transactions take the form of barter or other semi-legal payment arrangements (Huff, 2000; Stern, 1998; Commander and Mumssen, 1998). In the countries where electricity theft is widespread (such as Albania, Kosovo), or metering infrastructures are poorly-developed (Ukraine, Central Asian republics), energy utilities experience the most severe manifestation of non-payment: the existence of 'commercial losses': that is, 'non-billed consumption against which there exist no outstanding financial claims' (EBRD, 2001, p. 96). According to a World Bank (1999) study of the problem, 'Sector unbundling and privatization could generally help' reduce non-payment, 'but only if good corporate governance could be achieved, and strong mechanisms to enforce contracts and competent and independent regulation are in place' (p. 7). Otherwise, 'unbundling and half-hearted privatization could actually aggravate the problem, as can be seen from the experience of Russia, Armenia, Albania and Georgia' (ibid.). The study singles out Hungary, Bulgaria and Poland as positive examples of 'good corporate governance'; these countries have managed to solve the non-payment problem despite retaining state ownership or vertical integration of energy utilities.

The territorial variation of non-payment provides further evidence for the claim that energy poverty is closely related to the broader patterns of energy and economic reforms. In general, the problem is mainly concentrated in the states of the Former Soviet Union, and, to a lesser extent, some Balkan countries. Plotting the electricity bill collection rate (expressed as the percentage of payables within receivables) against the quotient of domestic to industrial prices during 2002 – the most recent year for which data was widely available – points to a paradoxical situation: the states which have the lowest relative residential tariffs tend to face the biggest non-payment problem (Figure 3.4). Most FSU countries – including Russia, Ukraine, Belarus and Azerbaijan – have decided to retain universal energy subsidies, which implies that non–payment should be attributed to institutional and regulatory factors beyond energy affordability. This argument is reinforced by the fact that politically unstable countries like Yugoslavia and Bosnia are also facing severe receivables problems, despite having rebalanced energy tariffs. Conversely, most ECE states are clustered above the 80 per cent collection rate and 1:1 tariff ratio, with the notable leadership of Slovenia, which inherited a Western-style price structure from Yugoslavia.

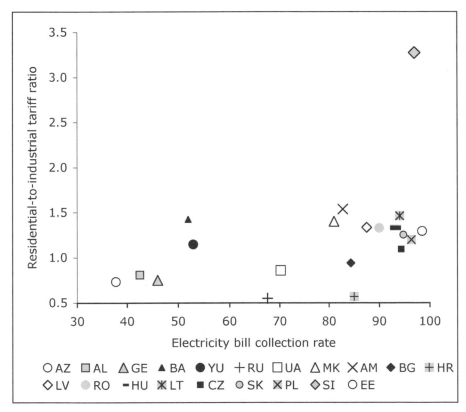

Figure 3.4   **Ratio between residential and industrial electricity prices, vs. electricity bill collection rate (percentage of payables within the sum of all receivables and non-billed energy consumption) in** 2002

Sources: IEA, 2003; EBRD, 2001.

Redistributive mechanisms for energy price rebalancing

Most post-socialist governments and energy utilities have endeavoured to provide some type of social support and consumer subsidy in response to the dismantling of previous energy pricing structures. Analysing the extent and nature of such schemes can contribute to an improved understanding of energy poverty factors, and their variation across countries. Current end-use energy subsidies generally fall into five groups:

Distorted residential energy prices   Keeping household energy tariffs below cost is still one of the most commonly-used social protection instruments, although it is being increasingly abandoned in the ECE region. Lovei et al. (2000) have found that 'most governments in the region maintain across-the-board utility price subsidies for political, rather than social reasons' (p. 24). Although these schemes are the easiest to administer in terms of transaction costs, it has been claimed that they exert a high fiscal burden on the state, while discouraging the efficient use of energy (Sobolewski and Zylicz, 2000; Kazakevicius et al., 1998). The 'lagging' transformation countries do not have the institutional capacity to replace this kind of support with more sophisticated schemes, which explains its persistence in the FSU and Balkan states with lower levels of institutional and economic development.

Below-cost pricing for energy is the least represented in the electricity industry, where pressure for commercialization has been the strongest. Natural gas and, particularly, district heating remain under-priced (and state owned) in many post-socialist countries. For example, although Hungary, Romania and Bulgaria operate fully market-based electricity sectors, the former two have chosen to retain implicit consumer subsidies for natural gas, while DH in the latter two continues to be under-priced (Velody et al., 2003).

Targeted price subsidies   Other than providing a universal price subsidy, social protection may also be embedded in the energy tariff by offering a reduced price for certain amounts or types of energy consumption. One of the most commonly used instruments is the 'lifeline tariff', whereby a basic amount of electricity is provided at a lower-than-average price, with all consecutive Kilowatt hours being charged at or above cost. The introduction of such mechanisms is contingent on the instalment of energy meters, which has hampered their application in many FSU countries, especially in the DH sector. Electricity lifeline tariffs are the most common throughout the post-socialist space, although, for example, Romania and Bulgaria have lifeline tariffs for gas. Serbia has achieved significant results by using such an approach, as the introduction of a three-block tariff system in this country:

> created an incentive for households to use electricity efficiently (and in particular to reduce the consumption of electricity for heating), while keeping the price of non-heating electricity consumption within the financial reach of most households. Between the winters of 2000/2001 and 2001/2002, this enabled EPS [the Serbian electricity company] to successfully reduce heating demand by 20 per cent, while in the same period, according to Household Survey data, household expenditure on electricity was a relatively modest 4-6 per cent of the household budget across all income deciles' (EBRD, 2003, p. 9).

Energy consumption may also be subsidized in the form of a 'social tariff' whereby energy is provided at below-cost to particular groups of customers. For example, Montenegro has developed a mechanism that enables welfare beneficiaries of 'Family Material Support' to qualify for a 23 per cent reduction in electricity bills (EBRD, 2003:9). The Albanian, Romanian and Bosnian governments have also introduced social tariffs to help particular groups of households to cope with price increases towards the end of the 1990s (EBRD, 2003). Many post-Soviet states (most notably Ukraine and Moldova) have opted to maintain the Communist system of merit-based price discounts for certain occupations (police, firemen, judges etc.), while providing the same kind of support for certain socially-vulnerable groups, such as low-income pensioners (Lovei *et al.*, 2000), although such schemes have gradually been dismantled.

Progressive pricing may also be provided in the form of a 'time of day' energy tariff, whereby the price of energy varies during the day. Such approaches contribute to the efficient use of energy, but are not always supported by the utilities, due to high meter installation costs and lower revenues. In some countries – such as Macedonia – they were removed after 1990 for precisely those reasons.

Non-disconnection of consumers in debts and arrears is common in some states of the former Soviet Union and the Balkans, where the energy sectors operate under political directives. The geography of this subsidy is directly dependent on the extent of state capture, energy privatization and metering: situations where the latter two factors are widely present lead to disconnection being focussed promptly on distinct groups of consumers (this is frequent in Central European states). Conversely, the use of non-disconnection as an economic subsidy is typical for many less-advanced states, such as Russia, Romania and Ukraine. This process usually alternates between two extreme states: first, households are allowed to use energy without paying, so that uncollected revenues often reach 20–30 per cent of billed revenues. However, when the utilities' financial and/or technical problems exceed political pressure from the government, the utilities have little choice other than mass disconnection. The inability to cut off individual consumers from district heating or electricity grids often means that entire villages, towns, and even regions, may be disconnected. The cycle reverts to the initial state when energy crisis-inspired public protest forces the government to supply 'emergency' budget financing to the utilities and affected population (see Burdett, 2002; Siner and Stern, 2001; Bradshaw and Kirkow, 1998; Timofeev et al., 1998).

Direct cash payments have been used by nearly all ECE and FSU states in the transformation period. Most often this has been implemented via targeted financial support to selected groups of disadvantaged households. One way of identifying the recipients of this aid is provided by the 'burden limit', which defines as a vulnerable household one whose utility expenditures exceed a given percentage of its total income. Burden limit-based financial support is usually administered via a network of social welfare offices which make payments to either utilities on the behalf of households, or the households themselves. Such schemes are common in the FSU: for example, Kazakhstan's social support system allows households whose total utility burdens exceed 30 per cent to qualify for energy assistance. However, Lovei et al. (2000) claim that the system of burden-based housing allowances reached only 36 per cent of income poor households in Ukraine in 1996. This is attributed to the absence of a strict disconnection regime, the high administrative costs of applying for the subsidy, and the relatively low level of monetary utility burdens among some low-income households. Such factors discourage many eligible families from applying for the allowance (see Dodonov et al., 2002; Brathwaite et al., 2000).

In most cases, governments base their financial support on the size of a household's earnings, relative to a given benchmark. For example, Bulgaria operates a 'Guaranteed Minimum Income' programme, which includes an energy assistance element for vulnerable households during the heating season. However, 'it is considered that current transfers are inadequate to meet basic needs' (EBRD, 2003, p. 6). A similar scheme in Romania allows vulnerable households to receive 'Heat Assistance Payments' which may be provided directly to the district heating and/or natural gas utilities, or to households heating their homes with wood, coal or bottled gas. The Moldovan Government has been offering targeted income support for 11 household categories deemed to be vulnerable. According to the IPA energy affordability study, this is a 'step up' from the previous system of Soviet-style 'social tariffs' for privileged groups of consumers (EBRD, 2003, p. 8).

Instead of having 'earmarked' payments – whereby particular groups of energy consumers are targeted by the state because they are deemed to be vulnerable – some countries have attempted to fight energy poverty through broader poverty alleviation programmes. This may be done by providing general allowances for utility costs, or by reinforcing existing social transfer systems. Such measures are easier to administer than targeted energy support schemes. They may be more efficient if the energy poor are a subset of the income poor, and the institutional network for detecting the income poor is well-developed (Freund and Wallich, 1996). A good example is provided by the Hungarian programme of housing allowances, which local governments have used to allocate national funds to a wide range of local needs, including energy poverty. Armenia also provides cash payments for utility-supplied heating based on the Poverty Family Benefit Programme, at around 14 US Dollars per household per month in 2002. This nominally includes a utility component and is paid in cash to some 25 per cent of households. However, because the payments are also supposed to cover other needs, the cash is often not applied to heating bills. The Czech Republic also offers housing allowances to households whose incomes fall below a certain threshold defined statistically by the state (Velody et al., 2003; Lux, 2000a).

In some cases, governments may provide one-off financial support for households struggling with energy payments. In addition to the programme described above, Hungary implemented a short-term energy assistance scheme that was designed to help the poor suffering from rapid price increases for residential gas and electricity. Financing was provided through the Hungarian Energy Fund, composed of funds from both government and industry. The programme was implemented in 1997 through 1998, to compensate households for extensive tariff rises during the previous year. In this case all recipients of social assistance were considered eligible, alongside a small number of 'non poor' households, pending formal approval from municipalities or social affairs offices (see IEA, 2000; Vorsatz, 1997). During the late 1990s, the

Czech Republic also operated a scheme that provided direct cash assistance for low-income households in selected regions where DH prices were increased at above-average rates during the late 1990s. Armenia offered cash payments for households who couldn't pay their electricity bills in 1999 and 2000.

It has been frequently pointed out that, in their present form, cash-based social safety networks for energy price increases provide inadequate protection for the poor. This may be due to either the insufficient size of the payments, poor targeting, or institutional obstacles in the cases where national social assistance payments are distributed by local governments. Velody et al. (2003) provide several examples of this situation:

> In Romania, local governments have been unable to fund social assistance benefits or have restricted access to social assistance benefits; the result has been that households receive smaller benefits or none at all. In Hungary, instead of spending national poverty prevention funds for their intended purpose, local governments often use the funds for other local needs. In both Bulgaria and Hungary, decentralization of the social protection responsibility has meant that cash-strapped local governments have been unable to provide sufficient assistance, thereby effectively decreasing the value and availability of social protection programs (p. 6).

These authors furthermore point out that there is a complete disconnection between energy efficiency financing schemes and social safety nets in the five post-socialist countries (Armenia, Bulgaria, Hungary, Kazakhstan and Romania) that they have surveyed. They note that none of these states have devised energy efficiency programmes for the poor, 'and none are seriously considering doing so'. The consequence of this situation is that 'most low-income households continue to have very limited abilities to improve the efficiency of their energy consumption in response to energy price increases' (ibid.).

In summary, it can be stated that social support mechanisms for energy price increases also have a specific pattern of spatial variation across the region. The EU accession states of ECE tend to use a narrower range of instruments, which are specifically targeted towards the income poor. Conversely, less-advanced reformers in the FSU region have resorted to more generic, *ad hoc* measures. Once again, this points to the importance of institutional and regulatory legacies in determining the nature and extent of social policy responses to energy sector reforms.

Spaces of energy poverty in post-socialism: towards a typology

The analysed evidence in this chapter points to the existence of numerous 'tapestries' of energy poverty factors and indicators in the states of the post-socialist transformation. The most important distinction arises with respect to the institutional patterns of energy restructuring, because the defining moment in the emergence of domestic energy non-affordability is the removal of universal price subsidies. The pace, nature and breadth of energy restructuring has also influenced the movement towards a more efficient energy economy across the transition space.

In this respect, von Hirschhausen and Wälde's (2001) analysis of the generic of patterns energy transformation provides a useful conceptual framework for situating the supranational geographies of energy reform. They distinguish among three 'stylised' outcomes of energy restructuring:

- The 'Post-Soviet mixed economy', comprising FSU states like Russia and Ukraine, which are 'a blend of incoherently functioning elements of a market economy and straightforward state planning' (p. 17);
- The 'Caspian state economy', which includes the Caspian FSU countries, characterized by 'an autocratic, clan-based regime based upon a strong state involvement in the economy' (ibid.);
- The 'Reforming Central/Eastern European market economy' – consisting of the respective ECE states which 'have largely adopted the formal institutions of a market economy or [have] at least given a binding commitment to do so in the future' (ibid.).

The reviewed sources indicate that the patterns of electricity reform, energy efficiency policies, practices of non-payment and social support mechanisms for energy price increases broadly conform to these patterns. Electricity reform – which is a good indicator of overall energy transformations – has followed a clear market-orientated path only in ECE countries, while FSU states possess a more mixed regime, where elements of the market regulation co-exist with remnants of socialism and the amalgamation of state and industry. These countries have been reluctant to raise energy prices to market levels, while maintaining implicit energy subsidies from industry to households and tolerating non-payment. Although energy efficiency policies in the residential sector remain under-developed throughout the post-socialist space, they are particularly weak in the FSU group, where the basic technical and institutional conditions (meters, correct price signals, independent regulatory bodies) are still lacking.

Table 3.1　　Total excess winter deaths (EWD) compared to annual mortality from transport accidents in selected European countries

Country	EWD	25% of EWD	Transport accidents
Armenia	2 000	500	1 100
Bulgaria	7 400	1 850	984
Germany	32 000	8 000	6 087
Lithuania	1 650	412	863
Poland	14 700	3 675	6 400
Portugal	9 000	2 250	1 760
Romania	17 600	4 400	3 270

Source: Bonnefoy and Sadeckas, 2006.

The patterns of social deprivation and exclusion in transformation constitute a second layer of variation in the underlying contingencies of energy poverty in transition.

Once again, there is a clear difference between poverty profiles in ECE countries, on the one hand, and FSU states, on the other. While East European poverty bears resemblances to 'Western' or 'developing world' patterns, where lower incomes tend to be correlated with the households' socio-economic status *vis-à-vis* factors like education, employment and family structure, the extent to which the demographic patterns of post-Soviet poverty conform to merit-based criteria is uncertain. This is despite the mounting body of evidence in favour of the argument that FSU countries are beginning to converge towards a similar social distribution of socio-economic vulnerability (Eatwell et al., 2000; Brathwaite et al., 2000). In statistical terms at least, they exhibit significantly higher levels of income inequality and lower degrees of human development than the countries further to the West.

One of the most interesting insights into the variation of energy poverty factors across the post-socialist space is provided by comparative data on 'excess winter deaths' (EWD): the seasonal increase in mortality over the winter period. Bonnefoy and Sadeckas (2006) established that even if a quarter of EWD were attributed to inadequately heated homes,

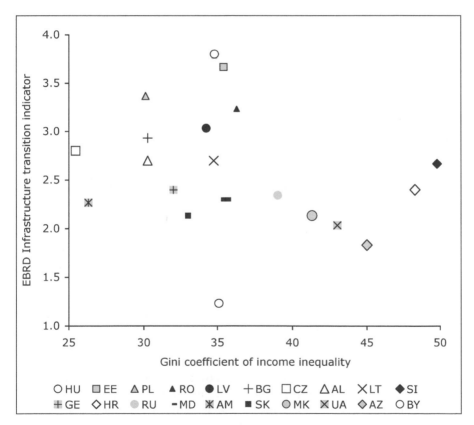

Figure 3.5   Relationship between average EBRD infrastructure transition indicators during 1995–2001 and average Gini coefficient of income inequality during the same period

Sources: Burdett, 2002; EBRD, 2001.

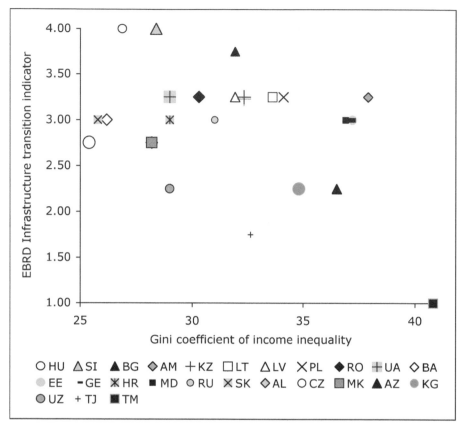

Figure 3.6   EBRD infrastructure transition indicators vs. Gini coefficient of income inequality in 2005

Sources: EBRD, 2005; UNDP, 2005.

this number is still higher than the annual mortality from transport accidents in most countries (see Table 3.1). However, although EWD are pronounced throughout the post-socialist space, their use as a cross-country comparative indicator remains controversial, because it is difficult to disaggregate the role of other environmental factors in excess winter mortality (see, for example, Healy, 2003), Still, it is clear that energy poverty plays at least some kind of a role in the emergence of this phenomenon.

The specificities of energy reform and poverty in some of the lesser-developed transformation states underline the importance of relying on a wider range of indicators in estimating the size of the energy poverty problem. Although the size of the non-payment problem is often a good indicator of domestic energy affordability *per se* (see Burdett, 2002; World Bank, 1999a) it should be used with caution in the case in many FSU states. This is because the driving forces of non-payment in such contexts are usually culture- and institution-, rather than poverty-related.

Comparing the EBRD's 'infrastructure indicators' to the Gini coefficients of income inequality across the post-socialist space (see Figures 3.5 and 3.6) indicates that the countries with high levels of income inequality also tend to have lower energy reform scores. (the values for the Pearson correlation coefficient are -0.20 and -0.30 respectively.)This suggests a link between relative income deprivation, on the one hand, and the extent of neoliberal energy reforms, on the other. It provides yet one more piece of evidence about the importance of institutions in the transformation process, because both income deprivation and energy reforms are closely dependent on the institutional strength of the state. Still, such correlations should be taken with a grain of salt, as both the Gini and the EBRD indicators are highly reductionist measures, and are valid for broad comparisons only.

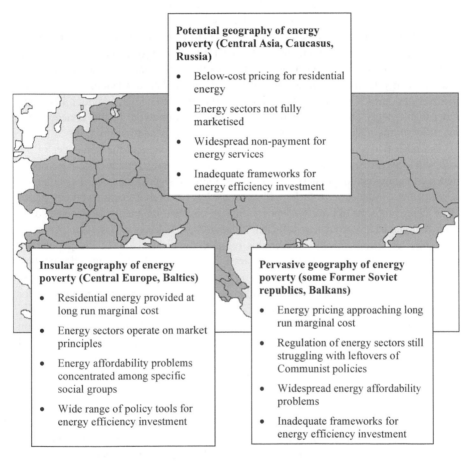

Potential geography of energy poverty (Central Asia, Caucasus, Russia)

- Below-cost pricing for residential energy

- Energy sectors not fully marketised

- Widespread non-payment for energy services

- Inadequate frameworks for energy efficiency investment

Insular geography of energy poverty (Central Europe, Baltics)

- Residential energy provided at long run marginal cost

- Energy sectors operate on market principles

- Energy affordability problems concentrated among specific social groups

- Wide range of policy tools for energy efficiency investment

Pervasive geography of energy poverty (some Former Soviet republics, Balkans)

- Energy pricing approaching long run marginal cost

- Regulation of energy sectors still struggling with leftovers of Communist policies

- Widespread energy affordability problems

- Inadequate frameworks for energy efficiency investment

Map 3.1   Generic features of energy poverty across the post-socialist space

The different spatial kaleidoscopes of energy reform, social welfare and poverty transformation in post-socialism can be generalized to fit into three broad models (see Map 3.1). The first of these would comprise the Central Asian states, Russia, and to a lesser degree, the FSU republics of the Caucasus, whose energy sectors have been classified within either the 'Post-Soviet mixed economy' or the 'Caspian state economy' by von Hirschhausen and Wälde (2001). The reluctance to move the energy sector towards a market-based framework is a defining moment in the appearance of energy poverty in these countries. It is not clear whether poor bill collection rates are due to the low affordability of domestic warmth, or whether they arise out of a combination of specific management practices and consumption cultures. The ratio of residential to industrial prices remains low. It is thus unlikely that non-payment can be explained solely on the basis of affordability trends, even though both levels of human development and income inequality rates are well above the remainder of the region. Still, masses of the population could fall into energy poverty if price increases are implemented without adequate social compensation or income-boosting economic growth.

It is for these reasons that I have decided to classify these states as possessing a 'potential geography of energy poverty'. Although the population has been frequently exposed to domestic energy deprivation – in the form of energy blackouts or mass-disconnection – these phenomena cannot be termed 'energy poverty'. As long as energy prices remain below-cost, and the receivables gap is not linked to affordability (that is, energy is understood as an entitlement rather than a service), the study of insufficient domestic warmth in these contexts will have to be placed under a heading wider than 'energy poverty', because the definitional framework of this concept does not allow for the inclusion of forms of deprivation not related to affordability.

The other two types of energy poverty are more tangible, and the remainder of the book will focus on them. They are represented by the 'reforming Central/Eastern European market economy', although there are variations within this generic model, depending on the particular combination of factors. If a country has been rebalancing its energy prices while facing problems of poverty, deprivation, non-payment for energy, inadequate social safety nets and insufficient energy efficient frameworks, it is likely to possess a 'pervasive geography' of energy poverty. Under this regime, significant parts of the population, including both low- and middle-income strata, would be suffering from affordability-related domestic energy deprivation. Based on the evidence reviewed in this chapter, it can be inferred that the Balkan states, plus Ukraine and Moldova, possess the preconditions for a 'pervasive' geography of energy poverty factors.

Conversely, an 'insular geography' is present in the 'advanced' transformation contexts, where economic growth, social protection, and the institutional structures of the energy sector have been sufficiently strong to offset increases in energy prices and poverty. Energy poverty in these countries is likely to be limited to a few demographic groups in a disadvantaged financial and housing position, much like the situation in the UK and other Western European countries. The finding that insular geographies exist in most Central European and Baltic states reinforces the conclusion reached at several points during this chapter, that the causal factors of energy poverty are linked to the broad reform environment of each country. The EU accession process is making a key difference in the reform of the energy and social welfare sectors.

Still, it should be pointed out that this classification can only provide a framework picture of the supranational distribution of energy poverty factors. Its purpose is to outline how the background conditions of domestic energy deprivation vary across different reform contexts. Any additional insights into the contingencies and inter-connections of such factors require further in-depth analyses, at a higher spatial resolution. The next chapter examines the extent and structure of domestic warmth deprivation in Macedonia and the Czech Republic, which are representative of two out of the three broad categories.

Chapter 4

The Institutional (Re)production of Inequality: Reconciling Energy, Welfare and Housing Reforms

The previous three chapters pointed to some of the ways in which the path of the post-socialist transformation process has varied across different countries, regions and cities. It is likely that such trends have also produced divergent patterns of energy poverty at the micro scale. However, further explorations of these dissimilarities require a closer look at the kinds of organizational and political dynamics that have shaped the evolution of energy, welfare and housing institutions. A key question in this context is: how does the emergence of domestic energy deprivation relate to decision-making patterns in the relevant policy domains?

As a result, this chapter concentrates on the institutional background of energy poverty in Macedonia and the Czech Republic. Its main objective is to examine the influence of broader economic and organizational processes on the development of domestic energy deprivation. Studying two post-socialist states with divergent development paths, it is hoped, can shed more light on the multiple linkages between territorial and institutional formations, economic change and social deprivation in post-socialism. Thus, the chapter aims to demonstrate how energy poverty may be linked to the inadequate co-ordination of energy, social welfare and housing policies. It focuses on the lack of properly targeted social assistance programmes, as well as the institutional inadequacy of support frameworks for energy efficiency in the domestic sector.

A comparative analysis of Macedonia and the Czech Republic can provide the building blocks for a deeper understanding of energy poverty in post-socialism. Cross-referencing the institutional contexts of energy, housing and social welfare policies in the two countries opens the space for connecting energy poverty with the path-dependencies of post-socialist transformation. This analytical opportunity stems from the political and economic discrepancies between the two countries: while the Czech Republic has managed to achieve developed-world status during the post-socialist transformation (as evidenced by its membership in the EU and OECD), Macedonia has faced numerous political and structural obstacles in implementing economic restructuring programmes, and commenced the EU accession process only recently. Still, in both states, the emergence of insufficiently warm homes is a direct product of the removal of socialist-era energy price subsidies which existed before 1990.

The chapter has three research aims. In the first instance, I examine how energy poverty is produced by the interaction of relevant decision-making institutions in the energy, social welfare and housing domains. This predisposes the chapter towards unravelling the organizational and political processes that may generate and sustain the geographies of energy poverty. Second, the analysis' comparative component is expected to provide further insight into the conduciveness of different institutional designs to energy poverty. To what end can national (and regional) variation be attributed to the institutional and economic legacies of the transformation process in the two countries? A third aim is to interpret the broad path-dependencies, interests and constraints that have shaped the relevant institutions' development over time. Having unravelled the institutional embeddedness of energy poverty, the purpose of this part of the text is to place the institutional production of domestic energy deprivation within a wider temporal and spatial context.

The background research for the chapter was undertaken on-site in Macedonia and the Czech Republic. It incorporated semi-structured interviews with policy-makers, professionals and households, in addition to a review of locally-published secondary literatures. A total of 45 government officials, company representatives and NGO activists were interviewed in both countries. I also had 10 personal communications for data-gathering purposes, and two interviews with senior decision-makers at the London Headquarters of the European Bank for Reconstruction and Development (EBRD). The interviews were held between August 2001 and December 2004. The gathered evidence was analysed with the aid of the theoretical tools outlined in the introductory chapter of this book.

The analytical framework of the chapter thus combines approaches taken from economic geography, institutional economics and political economy. Of particular importance are the notions – originating in institutional and evolutionary economics – of systemic capital, transformation costs and path-dependence, created by the non-convergence of economic systems, as well as the emergence of self-reinforcing economic mechanisms, such as the 'institutional trap' in post-socialism (Dallago, 1999; Polterovich, 1999; Yavlinsky and Braguinsky, 1994). Political economy models have also been employed, especially Buchanan's (1975) distinction between the building of an institutional infrastructure ('choice of rules') and the day-to-day policymaking process ('choice within rules').

Some parts of the analysis rely on von Hirschhausen and Wälde's (2001) identification of three political layers within the post-socialist transformation: formal institutions, informal institutions, and the 'societal consensus about the economic system' (p. 97). These concepts have been combined with strategic-relational state theory – in particular, Jessop's paradigm about the 'structural and strategic selectivity of the state' (Jessop, 2001; Hausner et al., 1995) – to develop a theoretical framework for interpreting the production of energy poverty within the context of the state's prioritization of a selected set of policies and social groups over others.

Despite the use of different theoretical approaches, however, the chapter nevertheless has an empirical focus, as its primary aim is to review and compare the operations of energy poverty-relevant institutions in the two countries. In other words, the above methodological approaches have been used only as a means of achieving an empirical objective, rather than contributing to the theories that devised them *per se*. Instead of applying a fixed theoretical matrix for the whole chapter, each analytical tool has been employed within a particular context, deemed most appropriate for its use. For example, the distinction between formalized and non-formalized institutions has been made when analysing the policy-making process in the energy, social and housing sectors, while the notions of transformation costs and non-convergence have been utilized in the analysis of the broad reform framework in the two countries. The theoretical models of the 'institutional trap' and 'structural selectivity' are introduced only at the end of the chapter, when drawing analytical conclusions from the described empirical evidence.

The following sections represent two parallel country-level studies of energy, social welfare and housing reforms in Macedonia and the Czech Republic. Each section commences with an outline of the wider economic and social histories of each country, in order to 'nest' the institutional analysis within a broader context. I then look at the regulation of energy, social protection and housing in each country. While it has been impossible to include a detailed assessment of all the operations in these domains, I have nevertheless attempted to execute a comprehensive, albeit brief, overview of the reform path undertaken by the relevant state institutions. The role of non-central state organizations operating at different scales is also considered. Yet the primary responsibility for making decisions about the affordability and efficiency of energy services to the poor rests within national-level institutions in the energy, social and housing domains; and the focal point of the analysis is on them.

Governance and energy poverty I: Macedonia

The Republic of Macedonia is a landlocked country of two million people, occupying 25,713 km^2 in the Central Balkans, between Greece, Bulgaria, Serbia and Albania (see Maps 1.1 and 1.2). It was a constituent Socialist Republic of the Yugoslav Federation before becoming an independent parliamentary democracy in 1991 (Stojmilov, 1995). The legacies of past economic policies are of particular importance for the emergence of energy poverty in this country. Even though Macedonia inherited an elaborate economic infrastructure from the former Yugoslavia, it faced a series of insurmountable political obstacles upon its declaration of independence in 1991. The escalating civil wars in the Balkans had a devastating impact on the economy, as large industrial enterprises – whose joint output constituted 42 per cent of GDP in 1989 – were forced to either scale down or completely terminate their operations. This resulted in massive redundancies, structural inefficiencies and the maintenance of socialist-era 'soft budget constraints'. Direct foreign investment, which could have ameliorated the difficult economic situation of the country, was discouraged by political instability (ESI, 2002).

The post-socialist economic downfall can be attributed, in part, to the state's inability to respond to a marketized economic environment, post-1992. Prior to being included in the Yugoslav Federation in 1945, Macedonia was an early industrial society, which had been ruled by the Ottoman Empire for five centuries before 1913 (the country was part of the Serbian kingdom between 1913 and 1941, having been divided between Bulgaria and Italy during World War 2). It was in a much more difficult situation compared to countries like the Czech Republic – or for that matter, the Yugoslav republics of Slovenia, Croatia and the Serbian province of Vojvodina – which were part of the more developed Austro-Hungarian empire.

In Macedonia, Yugoslav socialism took upon the task of creating an entirely new system of governance, while laying the framework for industrial development.

However, during the 1960s and 1970s, and aided by its position as the most liberal socialist state in ECE outside the Soviet sphere of influence, Yugoslavia began to achieve unprecedented levels of economic growth and capital accumulation (Szajkowski, 1999). Although Macedonia was traditionally an agricultural and sparsely populated country, it was engulfed by a rapid wave of industrialization and urbanization during this period. The country's GDP increased eightfold between 1950 and 1990, with the total population rising by 80 per cent up to 2 million, and the share of urban inhabitants doubling from approximately 30 per cent to 60 per cent (Stojmilov, 1995). Yet the Yugoslav socialist system was not entirely successful in 'smoothing out' the regional disparities inherited from the past. According to Jančar-Webster (1993):

> The Yugoslav variant of [Marxist-Leninist] ideology – socialist self-management – compounded the political rigidity of the one-party monopolistic state by creating regional equivalents: six republican one-party monopolies and local industry-government-party organizations ... The country may have been called Yugoslavia – but that country never had a common market, common economic development, or a common culture (p. 172).

Although this statement can be contested on a number of points, it is generally true that the southern parts of the former Yugoslavia – including Macedonia – were still struggling with developing-world problems well into the 1990s, unlike the Central European republics in the north. The Yugoslav south inherited a legacy of urban poverty ghettos, undeveloped rural areas with widespread subsistence agriculture, as well as heavy corruption within public-sector entities. However, it is also true that the former Yugoslavia prioritized the development of the industrial sector in Macedonia, as well as the development of fixed energy, transport and telecommunications infrastructures (MUPC, 2000; ISPO, 1982).

These conditions, alongside the existence of an autonomous public-sector apparatus at the state level, had a facilitating effect on the post-socialist transformation process. Macedonia's gaining of independence from Yugoslavia was relatively effortless, as the new central state quickly established itself in all areas of competence. The country's first democratically-elected government pledged to undertake a sequence of neoliberal-style reforms. This included an economic programme that prioritized the macroeconomic stabilization of the economy, by controlling the exchange rate of the currency and reducing inflation, which had reached record levels in the first year of independence (SSO, 1993).

Early restructuring efforts were constrained by two sets of external factors. First, the fall of the Iron curtain and the appearance of a new independent state in the central Balkans disrupted the delicate geopolitical balance of interests in the region. In particular, opposition within Greece reached critical levels. The Greek Government mounted an extensive campaign to block the international recognition of Macedonia, unless it complied with a set of political demands, such as changing its name and constitution. As result of this dispute Macedonia lost access to its key shipping routes via Greece, while existing as a non-recognized state until 1994 (Trajkovski, 2002).

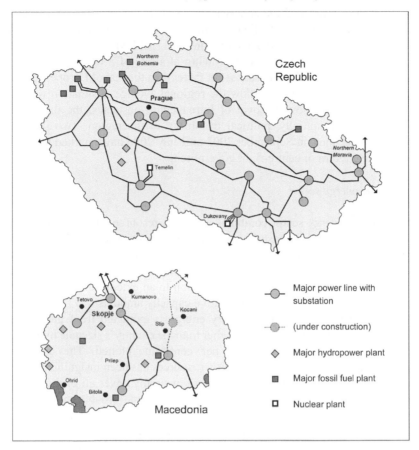

Map 4.1   Key electricity generation and transmission infrastructures in Macedonia and the Czech Republic

Sources: EPCM, 2003; USDoE, 2002.

Second, the escalating civil wars in the former Yugoslavia had a devastating impact on the Macedonian economy, already weakened by the Greek embargo and the post-socialist transformation process. The loss of markets in other Yugoslav republics implied that the majority of large industrial enterprises – whose joint output constituted 42 per cent of Macedonian GDP in 1989 – were forced to either scale down or completely terminate their operations, resulting in massive redundancies, structural inefficiencies, and the maintenance of socialist-era 'soft budget constraints'. The insufficient mobilization of human and fixed capital led to a decline in industrial output, while discouraging foreign investment (ESI, 2002).

Economic recovery ensued only in the late 1990s, but in an erratic manner. Although growth was facilitated by persistent macroeconomic stability (due to the firm control on economic policy exercised by the World Bank and IMF), political shocks continued to damage the economy. These included the 1999 Kosovo crisis and an armed insurgency of ethnic Albanians in 2001, which, despite being resolved peacefully, further worsened the county's foreign investment profile. GDP growth exceeded 5 per cent for the first time in 2000 (the year between the two military crises), only to dip to -1 per cent per annum in 2001 (MFRM, 2002: 23). Since then, the economy of the country has been growing at a rate of 3 to 4 per cent a year, slightly slower than other Balkan states, but faster than the rest of Europe. However, foreign investment has been far below the regional average, a situation which has been attributed to the slow removal of bureaucratic obstacles, the undeveloped banking sector, and, initially, political instability in the region. Two uneventful parliamentary elections, in September 2002 and July 2006 have marked the ousting and consecutive return of centre-right governments.

Reforming the energy sector

The Macedonian energy sector combines a modest resource base with a relatively well-developed supply and transmission infrastructure (see Map 4.1). The electricity generation system contains one heavy fuel oil- and four coal-burning plants (with a total installed capacity of 1010 MW), as well as 13 large- and medium-sized hydropower stations (458 MW total), in addition to an extensive network of small hydroelectric power plants, whose cumulative capacity amounts to 37 MW (MUPC, 2000). Nevertheless, most of the primary energy is extracted from the extensive (and highly polluting) lignite reserves, rather than the country's sizeable hydropower potential, which remains largely (80 per cent) unexploited. The commercial exploitation of geothermal and solar energy resources has been insignificant to date, although such sources are abundant and may potentially cover all of the country's energy needs. Fuelwood has emerged as one of the most affordable and widespread sources of energy for cooking and heating in the residential sector, particularly among low- and medium-income households (Stojmilov, 1995).

Unlike many developing states, Macedonia has an extensive energy distribution infrastructure: for instance, approximately 98 per cent of its households have access to electric power. The country consumes 7.1 TWh (25.6 PJ) of electricity per annum, 62 per cent (4.4 TWh–15.8 PJ) of which are used by households. Up to 6.2 TWh of electricity are generated domestically, while the remaining quantities are imported from neighbouring countries during periods of peak demand in winter (EPCM, 2003). Until recently, the entire electricity generation, transmission and distribution network was owned and managed by a single state-owned enterprise – the Electric Power Company of Macedonia (ESM, or *Elektrostopanstvo na Makedonija*). In 2005, this company was split up into separate distribution, transmission and generation entities, with a 90 per cent stake in the latter being sold to the Austrian copany EVN AG (*Energie Versorgung Niederösterreich*).

Other major corporate actors in the energy sector include the Toplifikacija Holding Company (a privately-owned operator of several fuel oil- and gas-powered district heating plants in the capital, whose total installed capacity of 478 MW constitutes the bulk of the country's district heating network), as well as GAMA – the joint stock company in charge of the limited gas supply system. The only oil refinery in the country and the retail petrol service network are both in private hands (AEA, 2006).

The energy sector is formally regulated by the 'Energy Act', which was adopted in 1997, as well as a number of lower-level legal provisions and decrees; including, for instance, the 'Methodology for Energy Pricing', which specifies the broad technical criteria for the setting of energy tariffs (MERM, 1998). Since 1997, there have been several amendments to the Energy Act:

> ... to allow for the creation of an independent regulatory body, that will issue licenses and set energy prices in a liberalized environment. The energy prices are currently set by a Government commission, and privately-owned energy companies are complaining that the process is untransparent and unfair (Interview with Violeta Keckarovska, Ministry of Economy, Skopje, 27 December 2001)

As a result of these changes, energy prices are now set by an independent body – the Energy Regulatory Commission of Macedonia. Its establishment predated the unbundling and privatization of the electricity industry, whose restructuring was partly motivated by the need to bring fresh capital and management. Though its problems were a far cry from those of similar utilities in the region, ESM was plagued by poor bill collection rates (from residential and industrial consumers alike), capital starvation and internal overemployment (MUPC, 2000). In addition, the development of the company was hampered by the interference of political elites and vested business interests in its management. ESM's predicament was further exacerbated by the challenging external environment faced by Macedonia during its 10 year existence as an independent country.

In light of these issues, it is easy to see why the commercialization and privatization of the electricity sector was a key priority of successive Macedonian governments, as well as the international financial institutions operating in the country. Most policy initiatives were concentrated on preparing the sector for the eventual entry of private capital (PEREEA, 2001; OECD/IEA, 1997). This necessitated significant energy price increases in order for electricity tariffs to reach the profit-making levels at which the power company would be attractive to commercial buyers. Even though tariffs were raised throughout the 1990s, and a World Bank assessment in 1998 concluded that an average electricity price between 4.2 and 5.4 US cents/kWh 'is probably not far from long run marginal costs' (LRMC) (World Bank, 1998, p. 11), the 2001 EBRD *Transition Report* estimated that the residential tariff of 5.3 USc/kWh constitutes only 70 per cent of the long-term marginal cost of this type of energy supply (EBRD, 2001:95). However, the government was reluctant to undertake further price reforms for fear of political resistance. The increase of tariffs by 13.5 per cent over a two-month period in the middle of the winter of 2000 was followed by mass protests by the Macedonian Consumers' Association:

> We called for a two-month boycott of electricity bill payments, because tariffs were increased very rapidly, and nobody could understand how the Electricity Company was setting the prices ... During our meeting with them it transpired that even their own officials could not explain how consumer prices were being determined ... We are against the premature privatization of the electricity monopoly, because that might introduce a foreign company with an even stricter disconnection policy (Interview with Kalčo Mitev, Macedonian Consumers' Association, Skopje, 19 September 2001).

Energy price increases were thus postponed for several years. However, this did not prevent the government from continuing to implement another one of its unstated policy objectives, also motivated by the desire to get a good bargaining price for the electricity utility: the heavy emphasis on investment in fixed assets. Throughout the 1990s, ESM continued to expand its infrastructural base, despite failing to operate on commercial principles. Thus, the early 1990s saw the construction of a new 250 km high-tension line between the main coal-burning station in the south (near Bitola) and the capital Skopje in the north. This project was financed with publicly guaranteed loans from international financial institutions, the repaying of which provided one of the main arguments for electricity price increases. ESM embarked on an even more ambitious investment programme in the late 1990s. This entailed the construction of several new hydropower stations, a 150 km high-tension line to Bulgaria, and two new high-voltage transformer substations:

> ESM would like to be the electricity transmission node of Southeastern Europe. This will allow electricity to be easily transported from Northern and Western to Southeastern Europe – Greece and Turkey in particular, where there is a high electricity demand over the summer, unlike the rest of Europe. In the more distant future, we could become a transit region for electricity exchanges between Europe and the Middle East and North Africa (Interview with Trajče Čerepnalkovski, ESM, Skopje, 26 September 2001)

Energy efficiency policies

The state's preoccupation with fixed capital investment and tariff re-balancing has been accompanied by the continued persistence (and in some cases, amplification) of the structural legacies of Communism. Even with a major movement away from heavy industry, the energy efficiency of the Macedonian actually worsened in the post-socialist period. According to the International Energy Agency, the energy intensity of GDP increased by 10 per cent between 1990 and 1998, with an accompanying growth in total CO_2 emissions from 10.24 to 10.83 Mtonnes/year, and a cumulative rise of 0.01 tonnes of CO_2 emitted per inhabitant over the same period (OECD, 2001). This trend is partly accounted for by the rapidly increasing share of (mostly thermally-generated) electricity within the country's final energy consumption, from 16.1 per cent in 1990 to 22.51 per cent 1998 (PEREEA, 2001, p. 5).

The worsening of the energy efficiency situation may be attributed, in part, to the extensive policy focus on energy supply expansion, rather than reductions in demand. The most pronounced example of this can be found in the *Plan for the Development of Energy Infrastructure up to 2020*, which was an integral part of the Draft Spatial Plan (see MUPC, 2000). It was almost completely preoccupied with the need for increasing the present energy supply, in order to 'meet growing demand in the future'. At the same time, opportunities for energy savings were ignored, and no reference was made to the wide range of efficiency strategies already devised – but not implemented – by the state. Even ESM, whose transmission and distribution network was bursting at the seams due to heightened seasonal and diurnal payloads, did very little to curb the rise in residential electricity use for heating purposes:

> Residential electricity use has been rising since the mid-1990s ... it has created numerous problems for our company, as we have to provide electricity in line with the needs of our customers ... At present ESM does not have any policies to facilitate fuel switching or to support insulation and/or the purchasing of more efficient appliances. They are a distant prospect (Interview with Jelica Trpčevska, ESM, 6 September 2001)

This situation transpired despite the existence of several state-supported programmes for industrial efficiency (see IEA, 1997; PEREEA, 2001). According to official information from the Government, such schemes led to the implementation of more than 100 projects to improve the energy efficiency of the capital stock within the industrial sector, thanks to financing from the state budget, as well as an 8 million credit line granted by the EBRD in 1994 (OECD/IEA, 1997, p. 370). The Ministry of Economy even claimed that the Programme resulted in 'an 8 per cent reduction in the total energy consumption on an annual level' (PEREEA, 2001, p. 7). However, this policy effort was not matched by the development of an adequate legal and institutional framework for the formulation and implementation of demand-side energy efficiency measures.

Without energy efficiency programmes to offset tariff increases in the residential sector, households have been left without any support for improving their domestic energy circumstances (see IEA, 1997; PEREEA, 2001). This situation is particularly detrimental for the thermal insulation of homes occupied by low-income households, who do not have access to housing credit from market-based institutions, while having their budgets stripped by rising energy prices. Such problems are supplemented by the absence of control mechanisms for monitoring the thermal efficiency of new housing (GTZ, 2001). According to a technical expert at the Skopje District Heating company,

> we are losing a lot of money due to the poor insulation of buildings. This problem has two aspects. Firstly, there is the issue of the low quality of the original construction work. The housing stock of the seventies and eighties – especially the eighties – is of good quality, because the standards in that period were very strict, in some cases stricter even than in Western countries. However, a problem of a different nature emerged in the late 1980s and has intensified during the transition ... very often the contractors will simply bribe the inspectors, who then give the building a seal of approval although it does meet the legal energy efficiency requirements (Interview with Ivica Sekovanik, Toplifikacija, Skopje, 27 September 2002).

The social support system

Social security and equity were the founding principles of the Yugoslav state, as every citizen was guaranteed employment, free health insurance, education and affordable access to basic necessities. In addition, some social groups were given special assistance and/or privileges, according to disability or 'special merit' criteria. This necessitated the development of an elaborate social policy apparatus within the governing framework of the state, which was retained by post-socialist Macedonia with a high degree of institutional resilience and constancy.

The post-socialist period has seen the adoption of several new regulatory documents in the Macedonian social welfare sector, although a formal reform strategy was never formulated (MLSP, 2005). These legal acts have led to the simplification of the cumbersome benefits system inherited from the pre-1990 era, improved targeting mechanisms, and a stronger institutional framework for the social system (Hutton et al., 2000; Braber, 2000). But social policies have also borne the brunt of the country's extensive poverty and inequality problems, because successive governments have remained committed to maintaining universal welfare support, despite fiscal constraints. As a result, the number of households in receipt of social assistance increased from 4500 in 1990 to 62,500 in 2004 (this is nearly 20 per cent of all households), while budgetary expenditure on social protection has surpassed the 100 million Euro mark in recent years (HCHRRM, 2002; MLSP, 2005).

However, universal welfare support is insufficient to address the difficulties of the energy poor, because it can't resolve many of the root causes of the problem, such as inefficient homes and inadequate heating infrastructures (for a wider discussion, see Boardman, 1991). Macedonia has not developed a targeted energy poverty-amelioration policy during the transition. Prior to the privatization of ESM, the only mechanism of this type was implemented by the utility itself, in the form of a relaxed disconnection policy. ESM allowed residential consumers to continue using electricity or district heat despite months of non-payment, without a predictable notice of disconnection. Also, non-paying consumers who were recipients of social benefits could split their debt into several instalments. But even with massive discounts, the debts to the utility easily exceeded the meagre monthly cash social assistance that was often these families' only source of income:

> My debt towards the Electricity Company is 18,000 denars. They told me that I can split it in 10 monthly payments. I am supposed to pay the first rate now, but I do not have any money. I feed my four-member family with 1,600 denars of social aid, and I should pay 1,800 for the electricity. I cannot afford to buy food, let alone repay my debts. I begged them to forgive part of the debt and not to disconnect my electricity, but to no avail. I should also pay for their court expenses, while I have no money even for the basic debt. … My children will freeze. They can sue me again. Maybe if I'm lucky they'll put me in jail, at least it's warmer there (Personal communication with chronic debtor, 3 December 2001)

With the privatization of the electricity company, the future of these practices remains unclear. They have not been replaced by any other social safety nets.

The district heating companies also allow some users to consume energy despite months of non-payment, but without a predictable notice of disconnection. *Toplifikacija*'s aggressive disconnection policies during the 1990s were especially softened during the winters of 2000 and 2001, when the non-payment problem was the most pervasive. This led of a decrease of consumer debts by 70 per cent:

> The disconnection procedure for district heat is complex. Basically we try to disconnect as few households as we can, because it is technically difficult and creates problems for the other users … the problem has become overwhelming in the past few years, so that if we were to disconnect all the households which we are legally allowed to, we would be

without 1/5 of our customers ... This is why we changed the disconnection procedure in 2000. Instead of taking non-payers directly to court we first send our staff to their home, to discuss why the debt problem emerged in the first place, and what can be done to resolve it without going to court. If the family is able to re-pay the debt in some way (for instance in monthly rates) then the story ends there. Otherwise we usually get a court order for disconnection. Then the whole pre-court consultation procedure is repeated again. We only cut off the heating supply to chronic non-payers (Interview with Ivica Sekovanik, 27 September 2002).

Even though the state has failed to include energy poverty assistance in the social protection system, interviews with social workers in the Municipal Centre for Social Protection of the City of Skopje indicated that there is an informal awareness of the issue in social policy practices. These professionals communicate with social assistance recipients in their districts on a day-to-day basis, with the aim of providing them with the basic knowledge and skills for improving their housing conditions:

> Most of the families who I work with are unable to afford paying for their heating. This is a result of their poor income and housing conditions ... Low-income families have found various ways to cope with the problem. Usually they only heat one small room, using fuelwood. Their energy needs are not very big anyway ... We have brought the problem to the attention of decision-makers in the regional branch of our Ministry, which controls our Centre, but it is all very politicized ... Policy-making officials in the central Ministry are only interested in their political careers, rather than making an impact (Interview with a social worker in Skopje, 17 December 2004).

The lack of state-sponsored energy poverty support was also acknowledged by a mid-level decision-maker in the Ministry for Labour and Social Policy:

> There is a need to find some kind of mechanism to address the issue of unaffordable energy among low-income households. The ESM and Toplifikacija schemes are simply insufficient. Too many households remain without protection (interview with Penka Nikolovska, Ministry for Labour and Social Policy, 4 October 2001).

Still, Macedonian social policy lacks a term for 'energy poverty'. Even though social sector professionals recognize the energy affordability problem, they see it as an income issue only. There is little understanding and awareness of the role of the housing stock on the price and quality of energy services.

Housing

The extensive ideological focus on housing production – rather than maintenance – may have contributed to the complete absence of state-sponsored technical and financial support for residential energy efficiency among low-income households. The only social housing programme implemented by the Macedonian government has involved the construction of new housing and subsequent rent regulation, but without a clear energy component. However, the residential energy efficiency issue has received some attention from the German government, which financed a bilateral technical co-operation project titled 'Energy Efficiency in Buildings'. Having concluded that 'the needs for energy in Macedonian buildings are greater than those in Germany, despite the similar weather conditions in winter' (GTC, 2001), the project was aimed at the harmonization of Macedonian energy efficiency standards with EU legislation, while financing demonstration projects for the application of new energy saving technologies. One of the pilot energy-saving schemes within this initiative was linked to the State's 'Programme for Social Apartments'. The project also formulated a new thermal efficiency code for buildings, in line with EU regulations (Interview with Milka Petkovska, GTC Energy Efficiency Project, 20 December 2001)

The reduction of thermal losses in buildings is all the more urgent in light of the structure of the building stock in Macedonia. Almost half of the country's inhabitants are concentrated in four key cities in the north (including the capital Skopje), while the remaining 35 per cent are distributed in and around seven urban centres in the remainder of the country (ibid.). Thus, although officially 40 per cent of the total population is rural, effectively more than 30 per cent of this number consists of inhabitants of peri-urban or suburban settlements, which were constructed in a semi-speculative manner during the rapid urbanization of the 1960s. Although they were gradually included in relevant planning documents, state housing policy has paid little attention to these districts, choosing to focus on the modernization of the residential stock in central and inner-city areas instead. Larger Macedonian cities thus contain a characteristic housing pattern, consisting of a mix of pre-war, high-density buildings in the urban core, surrounded by socialist-era apartment buildings in the inner city. These are supplemented by settlements of family homes in the suburbs and outer parts of the city, where housing quality may be substandard in the case of homes built outside the planning system (MUPC, 2000). Such suburban homes rank among the worst offenders in residential energy efficiency terns, as, in them, thermal losses through the building fabric are significantly higher than the prescribed code due to poor construction quality.

Governance and energy poverty II: Czech Republic

Although the Czech Republic was a centrally-planned socialist state until as recently as 1989, it now has a dynamic market economy, with high levels of investment and growth. The rapidity of this change has been facilitated by the inter-war history of the former Czechoslovakia, which had one of the most developed industrial economies in Central Europe (Matějů and Kreidl, 2000). It was also the only East European parliamentary democracy between 1918 and 1938. However, the country was subsequently occupied by Nazi Germany, only to become a Soviet satellite in 1948 (Machonin, 2000). During the Communist era, the Czechoslovak economy was closely controlled by Moscow, especially after the failed Prague uprising in 1969. The application of the standard set of socialist policy instruments (emphasis on heavy industry, distorted economic and social policies) had an overpowering effect on the market structures inherited from the country's previous economic system. Despite having existed within a developed industrial economy for 300 years, and as a marketized democracy for a further 30, the Czech Republic still had to cope with all the typical legacies of socialism at the end of the 1980s (Mlčoch, 2000; Kabele and Potuček, 1995).

This county's first non-Communist cabinet took office at the end of 1989, under the leadership of Václav Havel, a renowned playwright and human rights activist. At the time it was still part of the joint Czech and Slovak federal republic, which was, however, dissolved on the 23rd of July 1992, in the face of political disagreement among parliamentary parties (Potůček, 1999). The same year also saw the victory of the right-wing Civic Democratic Party (ODS) at the general election, in a lightning political success often ascribed to the populist approach of its leader, Finance Minister Václav Klaus. His government pursued extensive market-orientated reforms, designed to reduce the role of the state by supporting private initiative and entrepreneurship. Klaus's policies led to a distinctive 'voucher'-based privatization process, which was implemented at a significantly faster pace compared to neighbouring post-socialist states (McMaster, 2001, p. 4). Klaus' policies initially led to economic growth and increasing income inequality, with the dominant perception among the population being one of improved economic conditions (Matějů, 2000; Kreidl, 2000). However, despite its re-election in 1996, the ODS administration eventually fell to mounting popular dissatisfaction with its restructuring programmes, and an economic recession in 1997 and 1998. The Czech Social Democratic Party (ČSSD) managed to form a minority government, which pursued a more gradual reform approach, while emphasizing its commitment to the welfare state.

There was widespread disagreement among experts and the public alike, as to the pace and nature of the Czech Republic's neoliberalization process in the initial years of the transformation. It has been claimed that the 1997 crisis was produced by the 'hidden vulnerabilities' of the Czech economic system, created by the ODS's *laissez-faire* reform approach. Potůček (1999) found that 'the most important problems were the unstable and poorly functioning legal and institutional frameworks for privatization, unclear ownership relationships, and slow and inadequate financial restructuring and modernization' (p. 31). Other experts, however, maintained that 'until the 1997 economic crisis, domestic and foreign economists more or less agreed that the Czech economy had fared the best; it had succeeded in overcoming transformation problems to the extent that Czech leaders had a penchant for talking about the completion of the transformation process' (Adam, 1999, p. 50).

The divergence between these two sets of opinions mirrors the deeper ideological conflict between the advocates of neoliberal-style reforms, on the one hand, and the proponents of a more gradual, welfare-oriented reform approach, on the other. Although this struggle was prominently expressed via the divergence between ODS and ČSSD policies, Czech society is also interlaced with deeper tensions between the elements of new and old regulations. The country's governing system is a hybrid mixture of different legal and economic systems, as the legacies of socialism continue to persist within post-socialist spaces and institutions (see, for instance, Smith and Swain, 1998; Mc Dermott, 1997). At the same time, it has become apparent that the introduction of the free-market reforms has resulted in the multiplication of typical post-socialist features, rather than the replication of 'pure' neoliberal patterns of functioning (ibid.).

Reforming the energy sector

The Czech Republic's advanced electricity supply infrastructure stems from the industrial policies of the Austro-Hungarian era. With an installed generation capacity of more than 15 thousand TW, and almost 3000 km of high-voltage power lines (1/6 of which are double), this country is an electricity 'bridge' between Eastern and Western Europe (see Map 4.1). Czech electricity exports reached 12 TWh in 2000 and 2001 – which is nearly twice higher than the entire power consumption of Macedonia. The additional quantity of electricity (the Czech Republic exports 25 per cent of its total electricity production) is partly due to the recent activation of the environmentally-controversial nuclear plant Temelín, near the Austrian border (see Map 4.1, and PiEE, 2002: 5).

The implementation of post-socialist energy sector reforms has resulted in the formal unbundling of the socialist-era electricity monopoly. The Czech Electricity Company (ČEZ, or *Česke energeticke zavody*) is now a 70 per cent state-owned joint-stock company, which operates 10 TW of generation capacity, consisting of ten fossil-fuelled stations, thirteen hydro stations, two nuclear plants, three wind farms and one solar power station. The remaining 5 TW of the electricity generation infrastructure is installed in 12 privately-owned fossil power stations, who sell their electricity directly to the high-voltage grid operator (ČEPS), a wholly-owned subsidiary of ČEZ. The distribution network is divided between eight joint-stock regional electricity companies (AEA, 2006).

Almost every larger town and village in the Czech Republic possesses its own district heating network (formerly mostly coal- and oil-fired but increasingly converting to gas in the recent past). The total annual household energy consumption of district heat has reached the 80 thousand TW mark in recent years (AEA, 2006). The gas network is equally well-developed, even though 98 per cent of this resource is being imported from Russia and Norway. While efforts have been made to reduce the country's energy dependence on the Russian Federation – by building new gas and oil pipelines to Germany – Russian gas still out-supplies Western sources. In January 2002, the Czech government signed a contract to sell Transgas, the state gas company, to RWE Gas of Germany (USDoE, 2002).

The Czech Republic's energy sector challenges have been of a regulatory, rather than an infrastructural, nature. Initial attempts to reform energy operations were heavily influenced by the neoliberal ideology of Vaclav Klaus's centre-right government. The first 'Energy Policy of the Czech Republic' was approved in July 1992 (Decree No. 111/92), based on the antecedent 'Programme Declaration of the Government of the Czech Republic'. This document elaborated an ambitious plan for the quick introduction of market-based transformation measures, and the removal of the structural legacies of socialist rule (Kopačka, 2000). Its implementation was followed by the inclusion of the Czech Republic in a number of international bodies, such as the Union for the Co-ordination of Transmission of Electricity (in 1995), the OECD Energy Charter (1996), and the International Energy Agency (2001). This was a period of extensive restructuring and investment in the energy sector. Major undertakings included: reductions of the environmental impact of electricity generation by desulphurization and denitrification of coal-burning plants, liberalization of solid (1994) and liquid (1997) fuel prices, as well as the adoption of Act No. 222/1994, about the 'the business conditions and implementation of state administration in the energy branches and the State Energy Inspection Board' (GCR, 2001; Evans, 1995).

The overriding nominal objective of all subsequent regulatory documents in the energy sector was to 'prepare the energy economy of the Czech Republic for joining the European Union in every legislative and technical aspect' (Sojka, 2000: 221). Yet a perception developed during the late 1990s, that the restructuring process had gradually lost momentum, and that further progress was needed in terms of tariff rebalancing, liberalization and privatization (SEVEn, 2001; Kočenda and Cábelka, 1999). The new centre-left Government responded to such criticisms by formalizing its commitment to fundamental reform with the aid of a new 'Energy Policy'. The final version of this document was adopted in January 2000, having been formulated by the oft-criticized Ministry of Trade and Industry. The government then approved new energy legislation, aimed at 'marketizing' electricity and gas operations leading to the creation of a new Energy Regulatory Office, whose task was to 'set energy prices in a transparent and accountable manner', while 'building the framework for third party access to the power and gas networks' (interviews with Stanislav Trávníček and Rostislav Krejcár, Energy Regulatory Office, Prague, 15th April 2002).

Energy efficiency policies

The regulatory reforms of the late 1990s have resulted in the fragmentation of energy efficiency policies across several government departments. The list of organizations responsible for energy conservation measures and investment has included:

- the Czech Energy Agency, which implements the bulk of the 'State Programme of Support for Energy Saving and Use of Renewable Energy Sources';
- the State Environmental Fund, which operates within the Environment Ministry, providing a variety of small-scale grants, and interest-free or state-guaranteed loans, for demand-side efficiency investment;
- the Ministry of Regional Development, which controls the aspects of the 'Energy Saving Programme' pertaining to residential efficiency;
- the Czech State Inspectorate of Energy, which undertakes audits in the public sector and in subsidized enterprises, and provides training for energy managers and technicians.

The obvious omission from this list is the Energy Policy Department within the Ministry of Trade and Industry, which sees itself as the 'strategic core' of the energy sector, where 'matters of consequence' such as the future of coal mining and nuclear power have been decided, rather than 'soft' issues such as energy conservation (interviews with Jan Pouček and Vladimir Wilda, Ministry of Trade and Industry, Prague, 10 April 2002). However, the inadequate treatment of energy efficiency and renewables in the 2000 Energy Policy was criticized by the Environment Ministry's statutory Strategic Environmental Assessment of the policy:

The Ministry of the Environment expressed a negative opinion [about the MPO Energy Policy], claiming that [it did not] comply with adopted environmental policies. But this statement is not binding for the cabinet, which approved the new energy policy despite the negative opinion of the Environment Ministry (SEVEn, 2001: 33).

Yet it would be incorrect to state that energy efficiency has been under-financed in the Czech Republic. During the transition, this country established a range of well-funded energy savings programmes since the beginning of the transformation. This included 'general support for building insulation and information and control installations' as well as 'support aimed at selected demonstration projects' (Legro, 1998: 3). One of the most tangible results of such policies was the widespread renovation and retrofitting of apartment blocks built during the Communist era (see Plate 4.1). In addition, the 'Programme for Energy Saving Lighting' provided financial resources for the instalment of 156,500 new compact fluorescent lights, funded by ČEZ (CEEBIC, 1997). After 1997, most forms of energy efficiency support were unified into the 'State Programme of Support for Energy Saving and Use of Renewable Energy Sources' (IEA, 2002).

Still, the Centre for Energy Efficiency has estimated that subsidies for fossil fuels between 1994 and 1998 cost each Czech 661.9 US dollars, while the construction and operation of the two nuclear power plants amounted to 346.5 US dollars/citizen

Plate 4.1   These co-operatively owned apartment blocks at the Eastern outskirts of Prague are being renovated in order to reduce thermal losses through the built fabric. An extra layer of insulation has been added to the façade and windows have been replaced

Photograph by the author.

in central state support (SEVEn, 1999, p. 21). At the same time, energy efficiency support didn't even reach a tenth of the 'dirty' subsidies, at 6.1 US dollars per inhabitant (ibid.). This reflects the marginalization of energy efficiency among the government's priorities in the energy sector, although such levels of support in itself is significantly higher compared to other post-socialist countries (USAID, 2002).

The social support system

> Under the [Czechoslovak] socialist system the population was protected by comprehensive social programmes. With some exaggeration it can be said that they took care of people from the cradle to the grave (Adam, 1999: 126).

> Social security was guaranteed for everybody from birth in a state hospital to a state-supplied funeral, if necessary (Kuddo 1995: 56).

Similar to Macedonia, the Czech Republic inherited a well-developed social protection system from its former federal state. The basic principle of social policy during the command economy era was that society must provide for the welfare of people who cannot work due to old age, disability, sickness, or because they had not reached the working age (ibid.). Universal health care and education were also granted:

> The redistribution in the sphere of production in a way represented a sort of 'welfare contract' between the party-state and the population, which was marred only by its inefficiency and by the known, but not officially acknowledged privileges of the nomenklatura (de Deken, 1994, p. 138).

According to Potůček (1999, p. 70), the first phase of post-socialist social policy reform process (which coincides with the mandate of the first centre-right government) was characterized by 'a combination between social-neoliberal and social-democratic approaches'. This is reflected in the first formal policy statement for the sector – *The Scenario for Social Reforms* – which was based on the principles of 'active employment policy', 'liberalization and pluralization of the social insurance system', and the 'creation of a social safety net' (ibid.). Successive cabinets were reluctant to change this mix of market and socialist principles, possibly due to the traditionally high public status of social welfare professionals. According to Götting, 'altogether the Czech reformers' record seems the most successful so far, though even in this case the reform process is proceeding only slowly and is being carried out in a very pragmatic way' (1996:8). Czech researchers have reached a similar conclusion:

> The newly elected government in the second half of 1992, and subsequent governments, applied solutions resembling the political philosophy that dominated the pre-1992 period. This was due to the social and institutional path-dependencies from the past, as well as the length of the legislative cycle' (Potůček and Rodičová, 1998, p. 71)

The social welfare sector was subsequently organized into three components:

- The social insurance system, which was aimed at protecting citizens in a 'socially-predictable' disadvantaged position, such as old age, poverty, disability and unemployment. The basic principles and obligations of the state were established by law, although there were possibilities for additional *ad hoc* support schemes and the introduction of private capital.

- State social support, which provided various types of allowances and benefits for specific demographic groups, who would find themselves in 'socially-recognized' vulnerable circumstances (maternity, inadequate access to housing and transport services, and so on). This system was financed from the state budget. It contained both means-tested support, such as the child, transport, rent, heating and housing allowances, as well as indicator-based payments – family, military, birth and death allowances (Mácha and Woleková, 1998).

- The system of social assistance, which was 'residual in character'. It was aimed at supporting the households with incomes below the social minimum, or in 'social need' as defined by the Act no. 482/1991. These two groups were covered by a range of specialized social services, which aimed to improve their social situation over the medium- and long-terms (Průša, 2002).

The main social sector institution in the Czech Republic has been the Ministry for Labour and Social Affairs, which has also included the Department for Social Policy and Social Services, and the Department for Social Insurance and Income Policy. Compared to Macedonia, the Czech Ministry possesses a more robust organizational structure at the national level, and the provision of social services is generally more strictly controlled.

Unlike Macedonia, the Czech Republic has made several attempts to develop a specialized scheme to shield vulnerable families from energy and housing price increases. These efforts gave resulted in the establishment of a 'housing allowance' and a temporary 'heating allowance'. Although both payments have been relatively generous in monetary terms (the housing allowance reached ca. 15 per cent of the average wage in 2002), their eligibility and targeting criteria have been controversial. There is a widespread impression in the Czech sociological literature, that income-based poverty relief measures are inadequate to cover the increased price of energy services (Lux, 2001; Sunega, 2002). The experts' complaints can generally be grouped in two categories.

The first group of critiques refers to the fact that the income boundaries of social minima for different household numbers (which affects the size of benefits and allowances) are based on an equivalence scale that prioritizes the protection of families with children, in line with the historical roots of Czech social policy (Průša, 2000; Milková and Vašečka, 1998). But, while rapidly ascending income equivalence scales usually do a good job at reflecting the high additional costs of essential goods (such as food and clothing) per household member, they do not match the relatively low marginal cost of energy services with respect to increased household size (see, for instance, Wicks and Hutton, 1986).

Second, experts also disagree with the methods for disbursing the aid. The main eligibility criterion for the housing allowance is the size of the household's income relative to the social minimum; housing expenditure is not taken into consideration in any of the stages of the means-testing process. Yet running costs for energy and housing, which are dependent on the size and quality of tenure, are not necessarily related to the households' income status, due to the egalitarian nature of socialist housing allocation (Lux, 2000b: 37; Adam, 1996). Moreover, 'the allowances are based on a normative estimate of housing expenses', which is inadequate in 'a situation where there is a high spatial and social differentiation of rent and energy expenses':

> The current housing allowance system fails to meet its social objective and does not correspond to the models used in the EU ... This allowance is allocated with no regard to real housing costs. As a result, the housing allowance is more a part of a poverty relief programme than an effective instrument of housing policy (Lux, 2001: 197).

The senior policy makers who I interviewed in the Ministry for Labour and Social Affairs responded to such criticisms by stressing the detailed background work towards determining the size of the social minimum and the high transaction costs of administering a better-targeted allowance system:

> The monetary equivalent of the social minimum has been determined as a result of extensive research over a protracted period of time. We have determined that families with incomes up to 1.6 times the social minimum are deprived of adequate housing services, albeit this may vary depending on the household circumstances. But it is unlikely that deprivation would occur above this line... The Czech Republic does not have the budgetary resources to make radical reforms of the housing allowance system. Also, it has been judged that there is no political or social need to extend the heating and rent allowances that were effective until two years ago (Interview with Jarmila Škvrnová, Ministry for Labour and Social Affairs, 25 April 2002).

In 2006, the Czech Parliament adopted changes to the 'Law for State Social Support', with the aim of making the housing allowance more sensitive to spatial variation in housing costs. The new changes will be enacted in practice from the beginning of 2007, and will involve a gradation of housing allowances depending on the ownership of the dwelling and the size of the municipality that it is in. However, this system does not include any additional targeted measures to compensate households for energy cost increases. Just as in the Macedonian case, the Czech social policy system does not recognize the existence of a separate energy poverty problem.

Housing issues and policies

The Czech Republic opted against the direct housing liberalization and privatization policies enacted in neighbouring CEE countries, such as, for instance, the 'right to buy' approach (Bodnár, 1996; Struyk, 1996). Instead, successive governments implemented a more indirect policy, which started with a massive transfer of public housing ownership and management responsibilities from the state to local governments, alongside housing privatization, rapid withdrawal of the state from housing subsidies, as well as partial rent deregulation (Sýkora, 2003: 61). As a result of these changes, the country has witnessed a very slight decrease in public renting, while other forms of tenure, especially cooperatives, still comprise a significant proportion of the housing stock. Steinführer (2004) argues that the Czech Republic has in fact implemented a 'roundabout' privatization, which is bringing the country closer to the dualist housing model (p. 1).

Unlike Macedonia, the Czech Republic developed a formalized housing regulation framework at the very beginning of the transformation process. The first 'Housing Policy Concept' was passed by the Czech Government in 1991. It marked a radical break with socialist era legacies, foreseeing the complete withdrawal of the state from housing investment, and the creation of a real estate market. This is contrary to the pre-1990 situation, which was characterized by heavy state involvement in the construction and allocation of apartments (Sýkora, 1996; Michalovic, 1992; Musil, 1992; Telgarsky and Struyk, 1990). Still, as pointed out by Potůček (1999, p. 177), 'only one department, which consisted of approximately ten employees and did not include an economist until 1994' was entrusted with the task of housing policy formulation and implementation during most of the 1990s. He underlines that 'in the Czech Parliament there was no committee on housing. Had there been a head of the ministry or a parliamentary committee, more administrative and expert attention would have been focused on the issue' (ibid.).

The Czech state's low prioritization of housing among other restructuring issues was also detected by the Czech population:

> In response to the question of whether the government had a clear, long-term solution to housing problems, 55 per cent thought that this was not true, and only 12 per cent held the opposite opinion (Havelková and Valentová, 1998: 235).

It was only in the mid-1990s that the government decided to take a more active role in housing provision, by concentrating new responsibilities within the nascent Ministry for Regional Development. This entailed the development of a wide range of financial support schemes, with the aim of overcoming initial capital barriers towards real estate investment and maintenance, especially among low-income households. A specialized Housing Development Fund was created, with the relevant government departments in the finance, construction and social welfare being granted additional tasks in the construction and provision of housing infrastructures (MRD, 2001; Havelková and Valentová, 1998).

Although the state developed several programmes to support the improvement of thermal efficiency in buildings, far-reaching investment in the energy efficiency sphere is still hindered by the complex situation of the regulated rental sector. Rents during Communist rule were heavily subsidized and controlled by the state, with tenants being granted unprecedented privileges (such as the right to a temporally unlimited and inheritable contract of lease). Successive post-socialist governments have been reluctant to break the social status quo created by this situation, so that tenants' rights are still strongly protected and rents continue to be set by the state. Even the owners of newly-restituted or -privatized properties were obliged to uphold Communist-era lease contracts, subject to a rent cap in the majority of cases (de Deken, 1994; Telgarsky and Struyk, 1990). Although the rent ceiling has been increased by more than 15 times since the early 1990s, the regulated rents are still significantly lower than market levels in most cities (Lux, 2000a). The persistence of this 'under-reform pocket' has created serious inequalities and tensions in the Czech housing sector, to the extent that the Constitutional Court ruled in 2000 that 'the regulated system is unconstitutional because it discriminates against the right of ownership, and the right of property owners to receive financial compensation for their property' (Obadalová, 2000: 10).

While the wider ramifications of the rent control gridlock extend beyond the subject of this book, its relation to the energy and social sectors is relevant to the institutional production of energy poverty. The absence of a well-targeted compensation mechanism for energy tariff increases has induced the state to use rent control as an across-the-board social protection mechanism. The problem with this approach is that the combination of below-cost-recovery rent levels and distorted tenant-owner relations is highly unsustainable over the both medium and long-term, owing to its negative effects on both the maintenance (including energy efficiency) of the housing stock, as well as the spatial mobility of households (Kostelecký et al., 2002). The state's only remaining option, however, is to undertake fundamental cross-sectoral reforms with prohibitively high transformation costs. This conundrum resembles the generic features of the 'institutional' and 'under-reform' traps faced by less-advanced transformation countries (see, for instance, Hellman, 1998; Johnson et al., 1997).

In 2006, the Czech Republic adopted a set of new laws that will allow regulated rents to gradually rise to a pre-defined 'target' commercial level within the next 5 years, with the aim of escaping such reform traps (see Table 4.1). However, experts have critiqued the methodological and informational basis for the calculation of target levels, among other points (Boušová, 2006).

Table 4.1   Growth of regulated rents according to new legislation proposed in the Czech Republic

Town	Maximum regulated rent in Euro/m²/month	Target rent in Euro/m²/month	Average annual growth of rents, in %
Prague- Vinohrady	1.31	3.18	24
Ústí nad Labem	0.65	0.88	8
Brno – Bohunice	0.97	1.65	14
Hradec Králové	0.72	1.60	22
Zlín	0.70	1.20	14
Olomouc	0.91	1.44	12

Source: Boušová, 2006.

Conclusion

There were few common socio-economic trends in the territories of the present-day Macedonian and Czech Republics prior to the socialist era. Although these differences were, to a certain extent, reproduced in the post-socialist restructuring process, the reviewed evidence points to a number of shared structural problems in the operations of governmental, civil and private institutions, in Macedonia and the

Czech Republic alike. Both Macedonia and the Czech Republic have failed to provide adequate social protection and energy efficiency support in response to energy price liberalization. This indicates that 'lagging' and 'advanced' post-socialist states alike lack a sufficient understanding of the regulatory acts needed to achieve the desired policy goals *vis-à-vis* electricity sector transformation.

The institutional embeddedness of energy poverty

The low affordability of energy services can be attributed to the institutional cultures, interests and relations of decision-making bodies in the energy, social and housing domains. Jessop's (2001) theory about the 'strategic selectivity' of the state provides a useful pathway for understanding some of these dynamics. In both countries, the interviews showed that the *de facto* policy mix leads to a decreased affordability of energy services among low-income households. A strategic-relational approach would view the growing gap between energy prices and stagnant incomes in Macedonia as the product of a strategically-modified 'structured coherence', functioning at the national level. The recursively selected strategies and tactics – mobilized within a broader political attempt to introduce a neo-liberal regulation – have acted upon the structural selectivities of the state. They have promoted quick energy price liberalization, at the expense of adequate social protection and demand-side energy efficiency support.

A further contributing factor may have been the institutional cultures and knowledges inherited from the socialist era, when the social welfare programmes did not include an energy efficiency component, although post-socialist years have seen a strong focus on the promotion of technical and structural solutions for the supply side of the energy sector. The persistence of socialist mentalities is illustrated by the fact that the only subsidy provided in response to energy price increases is administered by the utilities themselves. This is a typical continuation of pre-1990 legacies, because enterprises acted as universal welfare providers during socialism.

A similar situation can be found in the Czech Republic, although the structural bias is more complex and subtle in its case. This is illustrated by the comparatively greater weight of energy efficiency and housing investment among the priorities of the state, which has developed a range of formal instruments and policies to help low-income households overcome the initial lack of capital for domestic energy improvements. Still, the main informal political power has rested within the socialist-mentality-dominated Energy Policy Department of the MPO, which, as indicated by the interviews, neither understands nor wishes to prioritize the provision of energy services for the poor. Similarly, there is great resistance within the MPSV towards the incorporation of household energy cost criteria in the social assistance system. Institutional inertia in this domain is also hindering the reform of the rental system, which in turn affects the ability of households to invest in the energy efficiency of their homes.

Moreover, there appears to be a broad non-formal conflict of culture and interest between the nascent (and less influential) Ministries for Environment and Regional Development, on the one hand, and the well-established – and more conservative – Ministries for Industry and Social Affairs, on the other; the clear losers of which are socially-disadvantaged and -marginalized parts of the population, due to the stalled reforms in the social support system, and in the rental housing sector.

One of the weaknesses of the strategic selectivity approach is that it neglects the role of private, civil and governmental organizations functioning at different scales, as a result of its extensive focus on the central state. Still, the analysed evidence – from both the Czech Republic and Macedonia – indicates that non-state actors have played a relatively minor role in the production of energy poverty, due to their general organizational and economic weakness. The increasing pluralization of governance in both countries may present an opportunity to mobilize a wider base of knowledge and skills to address energy affordability issues.

The role of institutional processes in the production of energy poverty

The geographies of energy poverty in Macedonia and Czech Republic resemble the generic features of the 'institutional trap' in transformation. By definition, this is a cyclical paradox in which the unintended effects of poorly-planned market reforms are impeding the future implementation of the kinds of policies that were support to alleviate their structural causes in the first place. In an institutional trap, the misalignment of initial reform attempts leads to a rapid increase in institutional transformation costs, creating inefficient organizations that can remain stable beyond the original context in which they were created (Polterovich, 1999; Yavlinsky and Braguinsky, 1994; North, 1990).

The entire Macedonian energy sector operates within the confines of an institutional trap. Although formal restructuring measures have been adopted in line with international demands, their real-life implementation has proceeded in a slow, piecemeal fashion. This is because the main obsession of the governing elites has been to increase energy prices – in order to cash in on the sale of the electric utility – while preserving the political and financial benefits of distorted investment decisions. They have yet to realize that decreasing the energy intensity of the economy may bring greater long-term benefits than some of the supply-oriented policies that are being promoted at the moment. The existence of inefficient institutions within the social welfare and housing sectors has created a specific policy gridlock, which reproduces the systemic legacies of socialism, while creating new structural problems typical for the post-socialist period.

Conversely, the Czech institutional trap functions in a more constrained, 'hidden' institutional space, at the boundaries of the energy, social and rental housing sectors. The absence of a well-targeted compensation mechanism for energy tariff increases has induced the state to use rent control as an across-the-board social protection mechanism. But this solution cannot entirely resolve the structural legacies of socialism in the residential energy efficiency sector, as it does not contain incentives for households to invest in the quality of their housing. The lack of compensation mechanisms to address the structural issues related to energy price increases is also evident in the wider policies within the housing and social sectors, as most state housing subsidies are not targeted for energy efficiency improvements, while the housing allowance in the social protection sector amounts to a general poverty relief programme. Still, the state's only other option is to undertake fundamental cross-sectoral reforms, with prohibitively high institutional transformation costs. This results in the emergence of an institutional trap. Indeed, as pointed out by Potůček

(1999) 'The Czech government has often resorted to the pragmatic postponement of needed reforms, in the domains where quick liberalization could affect the interests of larger groups of the population' (p. 84).

A second institutional mechanism that promotes energy poverty is the 'energy efficiency gap', which is embedded in the policies of state and private organizations, as it is, by definition, a result of the absence of public policy corrections of market failures (Jaffe and Stavins, 1994: 804). Both Macedonia and the Czech Republic are facing this problem, although it is more extensive in Macedonia, due to the poor integration of energy efficiency technologies into households post-1990. Macedonian state and private institutions are almost completely unaware of the need to increase the efficiency of household energy use, which is gradually addressing the effect of higher-than-average efficiency standards and technologies inherited from the Yugoslav era.

The broader systemic context

The political processes that produce domestic energy deprivation can also be attributed to a combination of political economy-based factors. The reasons for the Czech Republic's stronger institutional framework *vis-à-vis* energy poverty may lie in the interaction of internal politics with the external context. Several powerful international institutions – including the EU and multinational corporations – have persistently acted in the direction of minimizing the role of local elites in the Czech restructuring process. This has increased the political breadth of the transformation process, while reducing the gap between formal commitments and non-formalized action. Macedonia has faced a challenging reform environment, with negligible external involvement aside from a few developing-country-lending IFIs. The confined nature of the Macedonian transformation process may have reinforced the position of rent-seeking individuals and organizations within the central state.

Still, the presence of institutional traps – such as the Czech rent deregulation conundrum – indicates that the legacies of antecedent economic systems have also played a role in the production of energy poverty. The intensity of the structural remnants of the past has varied across economic sectors and countries. The Yugoslav period left Macedonia with a well-developed legal and infrastructural framework in the social and housing domains, although this was only partially true in the energy sector. Such factors may have combined with the country's traditionally weak institutional cultures, to reinforce the strategic selectivity of the post-socialist state. The situation in the Czech Republic is inverse, as its socialist-era housing and social infrastructures were weaker, while the energy sector was highly developed. The country's existence as an advanced industrial capitalist state in the pre-socialist period may have facilitated the institutional switch towards a market economy in the 1990s. This combination of factors is evidently less conductive to the production of energy poverty.

Whichever explanatory approach is taken, it is clear that the policies that help promote energy poverty are a product of a complex interaction between the cultures and interests of different institutions in the energy, social and housing domains. The next chapter looks at the outcomes of these organizational relations, by investigating the size and structure of the domestic energy deprivation problem in the two study countries.

Chapter 5

Layers of Vulnerability: Towards a Socio-Demographic Profile of the Energy Poor

Having examined the different reform paths taken by Macedonia and the Czech Republic in reforming their energy, social welfare and housing sectors, it is now time to look at the results of the policies that have been implemented. This chapter investigates the socio-demographic extent of energy poverty among the populations of both countries. My main aim is to establish, within the constraints of the available evidence, which social groups are affected by domestic energy deprivation. I also look at the some of the demographic and spatial circumstances of vulnerable populations, with the aim of exploring the role of housing infrastructures and urban/rural socio-economic disparities in the production of energy poverty. This analysis relies on different strands of evidence sourced from both national statistical agencies and the secondary literature.

My main argument is that energy poverty patterns do not always conform to general poverty structures, as the number of households suffering from energy expenditure problems does not match official poverty lines defined by the state. This is because the amount of useful warmth in the home is determined by factors additional to income, such as the quality of the housing stock, heating systems, and daily occupancy patterns. Thus, aside from monetary support, energy poverty-amelioration policies must also address technology and housing stock issues.

Measuring poverty

The chapter operates with the 'consensual' and 'relational' understandings of poverty and deprivation explained in Chapter 1. The reliance on such a combination of theories has required the amalgamation of several methods, including analyses of income and expenditure based on welfare economic approaches. These frameworks stipulate that the extent of poverty among a given population can be determined in at least three different ways.

First, the 'absolute' method – which operates with a commodities-based approach (see Chapter 1) – states that a household can be considered 'poor' if its total earnings fail to reach a pre-determined minimum income. This theoretical understanding has a normative equivalent: the 'absolute poverty line' (APL), which is usually calculated by adding up the minimal amounts of money needed to satisfy a given set of pre-defined 'basic needs', for households of different sizes. The resulting monetary amount is then adjusted on a seasonal and annual basis, to reflect consumer price changes.

The Czech Republic possesses an APL, also known as the 'subsistence minimum', which is calculated through a combination of normative and empirical criteria. It has been devised 'by finding out, with the help of scientific methods, the rational nutritional standards, and on their basis other needs important to be able to maintain a minimum standard of living' (Adam, 1999, p. 162). The resulting monetary amount is 'a socially-recognized and legally-established minimal income boundary under which deprivation occurs' (CSO, 2005). The APL mainly serves as a basis for distributing social benefits and allowances.

Second, in the case of the 'relative' method, the poverty threshold is defined in a relational manner, in line with the 'utility' framework outlined in Chapter 1. In this case, the poverty line is seen a percentage of a higher income level, where deprivation is *not* supposed to occur. According to the most commonly-used standard, the relative poverty line (RPL) can be positioned at 50 per cent of the median income of the entire population (Jones and Revenga, 2000). Macedonia and the Czech Republic alike have RPLs which have been set by the state at, respectively, 70 per cent and 60 per cent of the level of median income established by nationwide surveys of household expenditure. Families whose incomes fall below this figure are considered 'poor' for statistical and social policy purposes. The Czech RPL has a higher level than the APL: in 2000, 3.4 per cent of Czech households had incomes under the subsistence minimum, as opposed to approximately 7 per cent under the RPL (CSO, 2005).

Third, the 'subjective' approach attempts to extend beyond the reductionism of income and/or utility criteria, by stressing the 'capabilities' dimension. It defines poverty on the basis on the principle that, basically, an individual is poor if 'he/she considers him/herself poor' (Matějů, 2000; Mlčoch, 2000). At the heart of this theory is the notion that 'life may be seen as consisting of a set of interrelated functionings, which in turn are composed of beings and doings' (Saith, 2001:38). The latter may include both physical elements – as in 'being adequately fed and sheltered' – as well as 'more complex social achievements, such as taking part in the life of the community, being able to appear in public without shame, and so on' (ibid.: 110).

I have combined all three approaches in quantifying and qualifying domestic energy deprivation, because absolute, relative, and subjective energy poverty information alike can be extrapolated from published poverty data in Macedonia and the Czech Republic. However, the key moment in defining and describing energy poverty is the affordability of the final energy service, because, simply, a household is energy-poor if it cannot afford to purchase the necessary amount of warmth in the home. Thus, one of the key methods for estimating the demographic size of energy poverty is the 'compensating variation', which quantifies 'the price a consumer would need to be paid ... to be just as well off after ... a change in prices of products the consumer might buy' (Economics Glossary, 2006). In this chapter, the compensating variation expresses the percentage by which a given household's income would have to rise in year y, in order for it to be able to purchase the amount of energy that it was buying in a previous year x, before energy prices were increased. The higher the compensating variation for a given group of households, the greater its loss of welfare during the period between x and y. Comparing the compensating variations for different income strata can help estimate of the size and type of populations affected by energy price increases.

Income deciles

Most of the statistical analyses in this chapter are based on the division of households into deciles of income. Data classified in this manner are provided in the publications of the statistical offices of Macedonia and the Czech Republic. The statistics for each decile are a weighted mean of all the households in that group: the mean income of the households in a given decile decreases as its number falls, with the tenth decile being the 'richest' and the first the 'poorest'. It is important to note that the two countries group households into deciles in different ways:

- In Macedonia, the decile boundaries are determined by dividing the income interval between the maximum and minimum earners into ten equal bands, so that the total number of households in each decile varies but the income intervals for each band are the same;
- The Czech Statistical Office obtains its deciles by splitting the total number of households into ten equal groups according to income, so that the intervals between the top and bottom earners in each decile are different, but the total number of households per decile is equal.

There are two main methods of classifying the population into deciles. The first approach categorizes households depending on their total, or 'aggregate', income, while the second one divides them into income bands according to the equivalent income per household member. This is obtained by dividing total household income with the number of 'equivalent consumption units' in each family, to account for the economies of scale achieved in household consumption. Macedonia and the Czech Republic use different methods for calculating consumption units, although they are both similar to the 'OECD modified' equivalence scale. This system assigns a value of 1 consumption unit to the household head, 0.5 to each additional adult member, and 0.3 to each child (Hagenaars, 1994; Atkinson et al., 1995).

Data sources and limitations

As I have already pointed out before, the inadequate political awareness about insufficiently heated homes has resulted in the absence of nationally-representative data about domestic energy deprivation. Neither Macedonia nor the Czech Republic have undertaken any direct, purpose-made surveys of energy poverty to date. I have thus been forced to rely on proxies, rather than direct information, to estimate the extent of insufficient domestic warmth in both countries. Even though a significant body of indirect evidence is available from the publications of the national statistical offices, energy utilities, and international organizations operating in ECE, none of these sources are sufficient to provide direct evidence about the existence of energy poverty. This is why the chapter draws from all the available sources of data, in addition to the outputs of fieldwork in Macedonia and the Czech Republic.

Some of the key information has been provided by the Macedonian Household Expenditure Survey (HES) and the Czech Family Budget Survey (FBS). The HES is executed by the Macedonian State Statistical Office on an annual basis, using a two-staged random sampling frame of 4200 households (SSO, 2005). It divides households into three groups, according to their principal economic activity: 'agricultural' (families whose entire income originates from commercial or subsistence farming), 'mixed' (part of the earnings are based on agricultural production), and 'non-agricultural' (the family's revenue is generated by activities in industry or services). As for the FBS, it has a sampling frame of 3000 households representative of the entire Czech population (CSO, 2005). The Czech Statistical Office has published its results annually throughout the post-socialist period, with income and expenditure data being disaggregated according to income deciles, quintiles and quartiles, as well as four key demographic categories (families headed by employed adults, self-employed adults, farmers and pensioners).

However, it isn't possible to compare the datasets of the two surveys on any point other than energy consumption, even though they both disaggregate data for equivalent and aggregate income deciles. This is because, as noted above, the HES categorizes households on the basis of source of income (agricultural vs. mixed and non-agricultural families), while the FBS operates with broader socio-demographic groups, such as 'pensioners' or 'families headed by employed adults'. Moreover, the surveys lack direct energy poverty information, because this problem has yet to be recognized and conceptualized by data-gathering agencies.

It should be pointed out that a second set of standardized instruments for judging the extent of domestic energy deprivation in both countries is provided by national surveys of subjective well-being. Such polls are usually undertaken on an annual basis by state statistical agencies in order to assess households' experiences and judgements of deprivation (see CSO, 2006; SSO, 2005). Within this chapter, they form an additional basis for assessing and comparing the patterns of energy poverty in the two countries.

The HES, FBS and surveys of subjective well-being are the only nationally-representative and mutually-comparable methods of assessing the social structures of energy poverty in the two countries. In order to obtain a better picture of the demographic nature and extent of domestic energy deprivation, it has been necessary to use a wider range of additional sources specific to each country. This is particularly true in the case of domestic energy efficiency, which is almost entirely absent from national statistical surveys in Macedonia. Thus, and in order to assess the relationship between energy poverty and energy efficiency in this context, I undertook a statistical survey of households in the Macedonian cities of Štip and Skopje. It has provided important information about the relationships between energy costs and housing circumstances, and the energy consumption patterns of different types of households.

Household survey

Aside from official national statistics, the empirical corpus of the chapter is based on two quota-based surveys of 200 households each, in the Skopje neighbourhood of Pripor, with a total population of ca. 12,000 inhabitants, and Isar (9,800 residents) in Štip. These two districts were selected so as to provide a 'transect' of housing types in the relevant urban areas. The survey was implemented with the aid of local interviewers during the winter of 2002/2003, and in late 2004. Its samples were both random and representative of the main housing and social structures in the relevant urban districts (as they were cross referenced with census data). Aside from investigating the households' subjective opinions about the level of warmth and insulation in their homes, the surveys also compiled data about the size and type of the dwellings, as well as the size of the families' annual energy expenditure. These indicators were interpolated into the 'cost-of-warmth' ratio, which expresses the amount of funds necessary to heat a square metre in the home.

Pripor and Isar alike include a wide range of housing and social structures since they extend from the semi-rural edges of the respective cities well into the inner urban fabric. But they are also different in a number of important ways, as apartment buildings in Pripor have district heating, while Isar is generally more urbanized. The other major difference between the two districts is that Isar is in a medium-sized town, while Pripor belongs to the broader Skopje metropolitan area. This is Macedonia's growth engine, with more than a third of the official national economy and a quarter of the population. Nevertheless, Skopje contains vast social inequalities due to the welfare effects of the post-socialist transformation, and the antecedent spatial contradictions created by Yugoslav modernization. It also comprises a diverse range of housing types, which are in differing states of repair depending on the

Plate 5.1   Skopje's southern neighbourhoods, which also include Pripor, combine a variety of residential buildings of different age

Photograph by the author.

Plate 5.2   The district of Isar in Štip lies on a hill overlooking the city centre. It mainly consists of individual family homes, although there is some apartment housing in the low-lying areas

Photograph by Slobodan Apostolov.

financial status of their owners (as the majority of housing is owner-occupied). Most residential buildings are connected to a well-developed district heating network – based on hot water produced in gas- and oil-burning plants – which provides the most affordable method of domestic heating. Households beyond the reach of this system generally rely on electric storage heating or fuelwood.

Pripor lies on a steep mountain slope followed by an elevated plateau, overlooking Skopje's city centre in the river basin below. Although there are some apartment blocks in the lower areas closer to the city centre, the majority of residential buildings in this quarter are represented by family homes built either during the urbanization of the 1960s, or the last 10 years (see Plate 5.1). The outlying parts of Pripor coincide with the broader urban fringe of Skopje, which is almost completely rural in both spatial and economic terms. Thus, most of the local residents derive their income from subsistence agriculture.

Isar, on the other hand, includes the western quarters of the inner city of Štip, as well as parts of the city centre itself (see Plate 5.2). Despite being of considerably smaller size, this town is similar to Skopje in that it also underwent a socialist-led industrialization and urbanization dynamic during the 1960s and 1970s (Stojmilov, 1995). But such trends have been reversed in post-socialism, due to the declining economic role of industry, at the expense of services and agriculture. The secondary sector is currently dominated by low-wage manufacturing, mostly in the textile sector. Although most of the city lies on hilly terrain – which has placed technical limits on the construction of multi-storey buildings – the low-lying areas near the centre and northern outskirts contain a wider variety of housing types. This includes individual family homes, tenement blocks from the 1950s and 1960s, and socialist housing estates. Moreover, Štip tends to have a more compact and older urban structure than most Macedonian cities.

Structure of the chapter

The chapter consists of two parallel analyses of the size and structure of energy poverty in Macedonia and the Czech Republic. The relevant information is examined through similar algorithms. I first review temporal changes in fuel poverty, inducing factors such as energy price rebalancing and the cost of domestic warmth. These are consequently compared with trends in the national energy balance, to determine how the given combination of macroeconomic policies and social conditions has influenced the residential fuel mix at the national scale, and whether a signature of energy poverty can be detected already at this level. I then look at the compensating variation for energy, and subjective perceptions of energy poverty. This is compared with information about the 'receivables gap' (that is, the size of the non-payment problem) in the residential electricity and district heating sectors. The conclusion of the chapter pinpoints the socio-economic groups vulnerable to energy poverty. I highlight the manner in which the demographic and housing circumstances of energy poor households differ from the rest of the population.

In essence, the chapter is a comparative study of two countries, aimed at showing how institutions and space may interact to produce two very different patterns of energy poverty in different geographical contexts. But it has been impossible to provide a complete degree of symmetry between the two studies, because of the inconsistencies between datasets, and the different types of issues faced by each state. Although the same analytical procedures have been applied in each case, different issues have received different priority. Comparatively less space has been devoted to the study of energy poverty in the Czech Republic, due to the significantly smaller size of its poverty problem. The two case studies thus emphasize the most important problems in the different countries, rather than following a strict comparative structure.

Energy poverty patterns I: Macedonia

The transformation of the energy sector in this country led to energy price increases exceeding 500 per cent in absolute terms since 1991, and approximately 150 per cent over the inflation rate during the same period (SSO, 1993). The effect of this price rise on the structure of residential energy use is evidenced, among other indicators, by the shift towards biomass (principally wood and wood products) in the national residential energy balance. The use of fuelwood – presently the most affordable and easily accessible means of obtaining domestic warmth – has risen by 44 per cent since 1991, so that almost 70 per cent of the population currently relies on it for domestic heating, especially in rural areas (although the share of households who

Energy Poverty in Eastern Europe

use it as the *main* heating source is somewhat lower, at around 60 per cent). In part, the increased use of fuelwood can be attributed to the expansion of illegal logging, as a result of the inadequate restructuring of state-owned forestry enterprises. The country has a black market for this resource, where it can be obtained at significantly discounted prices. A resent investigative report found that more than 3.5 million m³ of fuelwood have been obtained through illegal logging since 2000 (Pisarev, 2006).

District heating (DH) provides one of the most efficient and affordable sources of domestic warmth. According to *Toplifikacija* – the largest DH company in the country – this method is 15 per cent more efficient than wood, and 90 per cent more efficient than coal. However, although *Toplifikacija* heats 3.9 million m² of residential and office space in Skopje with hot water produced in five gas- and heavy fuel oil-burning plants, DH networks outside these areas are completely undeveloped, aside from limited systems in Kočani and Bitola. Thus, approximately 70 per cent of the population is without access to this type of fuel (Toplifikacija, 2005). Technical and economic models about the comparative costs of different heating modes in mean Macedonian conditions indicate that fuelwood and district heating provide the cheapest sources of domestic energy. However, the affordability of fuelwood depends on whether it has been purchased via illegal chanels, or through the official market (ibid.).

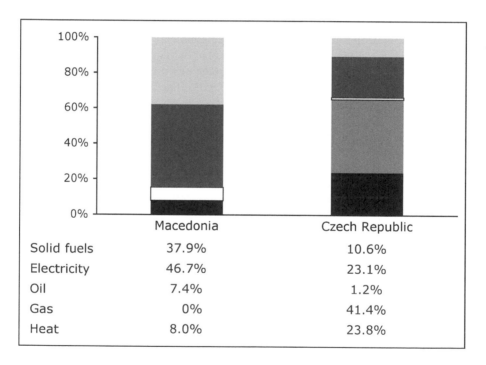

	Macedonia	Czech Republic
Solid fuels	37.9%	10.6%
Electricity	46.7%	23.1%
Oil	7.4%	1.2%
Gas	0%	41.4%
Heat	8.0%	23.8%

Figure 5.1   Structure of final residential energy consumption in Macedonia and the Czech Republic

Source: AEA, 2006.

Note: Percentages may not add up due to rounding.

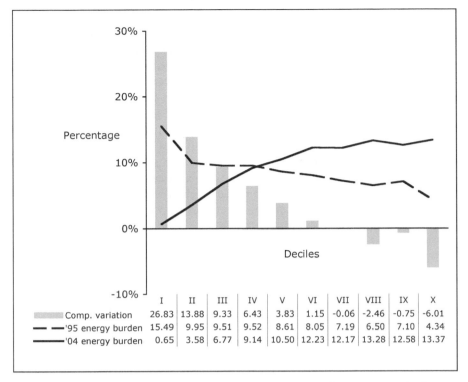

	I	II	III	IV	V	VI	VII	VIII	IX	X
Comp. variation	26.83	13.88	9.33	6.43	3.83	1.15	-0.06	-2.46	-0.75	-6.01
'95 energy burden	15.49	9.95	9.51	9.52	8.61	8.05	7.19	6.50	7.10	4.34
'04 energy burden	0.65	3.58	6.77	9.14	10.50	12.23	12.17	13.28	12.58	13.37

Figure 5.2   Compensating variation for energy expenditure per equivalent income decile, Macedonia, 1995–2004

Source: AEA, 2006.
Note: Author's calculations based on household expenditure data from SSO, 1996; 2005.

In the areas without DH where fuelwood is not available, households are forced to rely on electricity for heating. This is mainly the case in medium-sized towns containing large housing estates without access to district heat: Prilep, Štip, Tetovo, Ohrid, Kumanovo, and to a lesser extent, Bitola. Thus, almost 30 per cent of households in the country have been using electricity for heating, although it accounts for almost 50 per cent of the total fuel mix in the residential sector (see Figure 5.1, and AEA, 2006). The country's dependence on electricity for domestic warmth grew during the early 1990s, mainly because of artificially low prices left over from socialism, and the expansion of the unrecorded economy during this period. It has been accompanied by an increased sharpness of the daily demand curve in winter, causing technical problems for the electricity companies which are facing very high loads during the colder parts of the day. However, the use of this energy source has stagnated in recent years, as a result of rising residential electricity prices. In general, the energy costs of families who have to rely on either electricity or oil for heating are twice higher than those who use illegal fuelwood.

Energy poverty indicators

The emergence of energy poverty in Macedonia has transpired against the backdrop of a rapid increase in general poverty. According to a nationwide assessment carried out by the World Bank in 1999, as well as several similar studies by local institutions, Macedonia experienced a dramatic rise in income poverty during the early years of the post-socialist transformation (World Bank, 1999c; Hutton et al., 2000; ISPJR, 1998). The poverty growth rate slowed down in 1995 and 1996, only to be replaced by a second acceleration in the late 1990s (ibid.). As a result, the share of the population living under the RPL now stands at approximately 30 per cent, up from 4 per cent in 1991 (SSO, 2005). It has been established, however, that the poverty headcount is sensitive to the level at which the poverty line is set, because there is a statistical 'bunching' of households just below the poverty line, that is, poverty is 'shallow' (ibid. and Braber, 2000).

The change in the 'energy burden' – that is, the rise in the share of energy expenditure within the total household budget during the transition – provides interesting insights into the spread of domestic energy deprivation in Macedonia. According to the HES, in 1995, when implicit energy subsidies were still extensive, only the first equivalent income decile had an energy expenditure higher than 10 per cent (this is the 'cut off' point for energy poverty according to the mainstream literature – for example see Boardman, 1991). However, 2004 saw a reversal of the distribution of energy burdens across deciles. The top six deciles had expenditures higher than 10 per cent, while the bottom four actually spent much *lower* shares of their household incomes on energy (see Figure 5.2).

This is a paradoxical situation in standard welfare economic terms, as it would be expected that the poor would spend a *higher*, rather than lower, share of their income on energy. It can be explained by the extensive reliance on illegal fuelwood among the income poor, who either obtain this resource at a heavily discounted cost by avoiding official channels, or use it through subsistence forestry in the cases of rural settlements in woodland areas. Also, Macedonian households with lower equivalent incomes tend to be large extended families living in overcrowded housing, where the expenditure per household member may be very low. Household with high equivalent energy expenditures and burdens tend to be urban pensioners, who in this case are grouped in the higher income deciles as their pensions are received in monetary form, and are part of the formal economy.

For these reasons, it is better to rely on the compensating variation as a means of obtaining a statistical estimate of the size of the population affected by energy deprivation. This can be done by comparing the absolute energy expenditure of all Macedonian households in 2004[1], to the same figure in 1995. The grey columns in Figure 5.2 depicts the percentage by which incomes would have to change in 2004, in order for households to be able to retain the same ratio of energy expenditure relative to the national average in 1995. In other words, it shows whether income would have to be given to, or taken from, a household in order for it to retain its

---

1   This is the most recent year for which data standardized along the same methodology for 1995 was publicly available at the time of completion of this manuscript.

Table 5.1   Method for calculating the compensating variation in Macedonia

Decile	I	II	III	IV	V	VI	VII	VIII	IX	X	Average
Energy expenditure†	1,594	2,072	3,129	4,234	4,967	5,679	6,042	6,269	7,763	8,121	4,987
Energy expenditure ratio*††	0.32	0.42	0.63	0.85	1.00	1.14	1.21	1.26	1.56	1.63	1
Energy expenditure†	82	926	2,869	5,418	7,934	11,318	13,243	16,781	17,993	32,165	10,873
Expected energy expenditure**	3,475	4,517	6,822	9,231	10,829	12,382	13,173	13,668	16,925	17,706	10,873
Difference **	3,393	3,591	3,953	3,813	2,895	1,064	-70	-3,113	-1,068	-14,459	0
Total household expenditure	12,649	25,874	42,370	59,293	75,553	92,569	108,816	126,389	143,025	240,557	92,710
Comp. variation††	26.83%	13.88%	9.33%	6.43%	3.83%	1.15%	-0.06%	-2.46%	-0.75%	-6.01%	0%

(Rows 1–2: 1995; rows 3–7: 2004)

* Ratio of energy expenditure in each decile to average energy expenditure for all deciles.
** Based on 1995 expenditure ratio (this figure has been obtained by multiplying the actual energy expenditure in 2004 to the expenditure ratio for 1996).
*** Difference between expected and actual energy expenditure.
† In Macedonian Denars
†† Ratio between difference of expenditures and total household expenditure.

Source: SSO, 2005; 1996.

energy expenditure level relative to a normative value: in this case, the national average (see Table 5.1 for a more detailed explanation of the analytical procedure involved in obtaining the compensating variation).

It has transpired that the 60 per cent of households with lowest incomes, that is, those in the first 6 deciles, would have to receive additional funds – ranging between 27 to 1 per cent of total equivalent income – in order for the ratio of their energy expenditure to the national average to remain equal to the 1995 level. At the same time, however, income would have to be 'taken away' from the top 30 per cent in order for their ratios to remain the same. This means that the relative energy expenditures of better-off households have increased in comparison to the 1995 level. Such households have responded to energy price increases by allocating additional income for energy expenditure. The 60 per cent figure corresponds to the results of surveys of subjective well-being, according to which only 38 per cent of all Macedonian households thought that they were able to keep their home 'adequately warm' in 2003, although the same figure stood at 46 per cent only three years earlier (see Figure 5.3).

These figures are significant in three ways. First, they demonstrate that energy poverty has a much wider demographic extent than statistically defined income poverty, which does not include households above the third decile. This points to the inadequacy of the RPL defined by the state, which, it appears, has failed to include at least half of the households suffering from energy poverty. Second, the compensating variation shows that energy expenditures have become more polarized, because the top 30 per cent of households with highest incomes now have a greater energy expenditure compared to the 1995 level, while the bottom 60 per cent have been forced to cut back on their energy purchases. This leads to the third finding: that residential energy efficiency improvements have yet to be felt among the wealthiest parts of the population. In normal circumstances, their energy expenditures would be expected to decrease as a result of more efficient building installations and/or fuel switching.

Socio-demographic profiles of energy poverty

Although there has been no specific research aimed at identifying the demographic groups vulnerable to energy poverty, it is possible to synthesize some information based on the findings on the basis of other studies. For example, the HES indicates that energy poverty in Macedonia is disproportionately rural, because farming and mixed-income households are over-represented among the two lowest income deciles. According to the 1994 Census, 95 per cent of all farming households are rural, while 97 per cent of non-agricultural households live in cities (SSO, 1995). This trend is mirrored by the information gathered by the bill collection department of ESM (the national electricity monopoly), which indicates that arrears and debts in energy receivables are more pervasive in rural areas. Out of the 200 households interviewed in Štip, 47.9 per cent experienced non-payment difficulties.

The survey of subjective well-being within the HES showed that only 21 per cent of all farming households were able to heat their homes to an 'adequate' level in 2003, down from 44 per cent in 2000 (see Figure 5.3). A significantly above-average number of households in this group could not afford to pay their utility bills 'on time' – 49 per cent in 2003, compared to a national mean of 29 per cent. Although this group has access to non-monetary sources of incomes, which are deemed to have played a poverty-ameliorating role during the transformation – especially with respect to the use of fuelwood – this has clearly been insufficient to offset their low monetary incomes. The fact that most energy payments are requested in cash reduces in the weight of subsistence incomes.

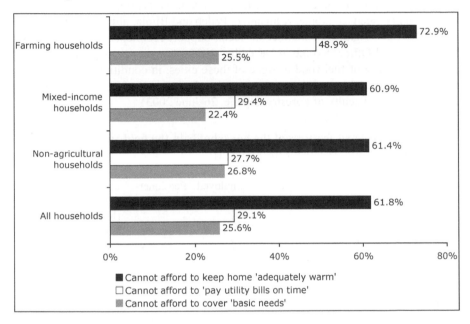

Figure 5.3   Subjective experiences of energy poverty, Macedonia

Source: SSO, 2005.

The hardship experienced by this group can also be attributed to the poor quality – and, hence, probably, the poor energy efficiency – of its dwellings. A survey carried out by the Skopje-based Institute for Sociological, Political and Juridical Research (ISPJR) found that 38 per cent of households living in mountainous villages thought that their primary dwelling was of 'unsatisfactory' quality, while the corresponding rates for villages in plains and urban areas were approximately 18 per cent in both cases (ISPJR, 2002). These findings match the outcomes of the 1994 Census and the background research for the National Poverty Reduction Strategy, which have established that substandard housing is overrepresented in mountainous villages. Such areas contain disproportionate number of low-income households, whose economic impoverishment and social marginality constrains energy efficiency investment (FCSRD, 2002).

A second vulnerable group is constituted by low-income households living in urban areas. The *National Poverty Strategy* points to the expansion of poverty among urban non-agricultural households with unemployed adults: 'the rate of poor people … in [the capital city] rose from 12.5 per cent in 1997 to 26.3 per cent in 1999, and the poverty gap for the same period, increased from 3.0 per cent to 8.3 per cent' (FCSRD, 2002, p. 12). The Strategy attributes these dynamics to a number of underlying processes, such as the high rate of unemployment, low pensions, and social segregation. A recent EBRD-sponsored study of electricity affordability in Macedonia listed welfare beneficiaries and the unemployed among the most vulnerable groups to electricity price increases (EBRD, 2003, p. 127). Such households tend to be concentrated in large cities in the central and southern parts of the country, such as Skopje, Kumanovo, Prilep and Bitola, although remote rural areas also have significant number of social support recipients (Interview at the State Statistical Office, Skopje, 9 December 2005). Local experts pointed out that poor accessibility of fuelwood in some of these cities, in conditions where district heating is not available, may have contributed to the increase of energy poverty rates (Interview at the Faculty of Forestry, Skopje, 26 June 2003).

Table 5.2   Selected features of the households in the field survey in Štip and Skopje (see also Buzar, 2007a)

Household type	All households	Single parents	Unemployed adults with young children	Pensioner-only	Extended families (living in a house)	Extended families (living in a flat)
Share of households living in 'insufficiently' heated homes	46.2%	48.1%	67.7%	50.4%	42.4%	50.1%
Monthly cost of heating 1 m² in the home, in Euro	0.352	0.370	0.675	0.527	0.281	0.461

Families with more than two children, and families with young children (under 7 years of age), also seem to facing above average rates of energy poverty-related hardship. The survey in Štip and Skopje indicated that there are nearly twelve times as many young children per working adult in the households with insufficiently heated homes, compared to those without energy poverty problems. Families with young children were facing the highest energy costs relative to other households: on average, the heating of each square metre in their homes required 0.67 Euro per month, against an average of 0.35 Euro/month for the entire survey sample (Table 5.2). The high energy expenses of such households can be attributed, in part, to the poor energy efficiency of the building fabric and heating systems in their homes. However, their above-average energy demands also play a role (for a wider discussion, see Wicks and Hutton, 1986). These findings are supported by the results of recent systematic poverty assessments by the World Bank (1999) and the ISPJR (1998) which point to the increasing incidence of poverty among families with young children. In general, such households are over-represented in northern and western Macedonia, in rural and urban regions alike.

The EBRD-sponsored IPA energy affordability study identified pensioners, especially those with low incomes, as a group vulnerable to energy price increases (EBRD 2003). This is further supported by the finding that the households with inadequately heated homes the Štip and Skopje survey had the highest relative number of pensioners (approximately 60 per cent of all the households in this group). The survey also established that pensioner-only households also had significantly above-average domestic energy costs, at approximately 0.53 Euro per square metre per month (Table 5.2). This figure matches the findings of national expenditure surveys, and may reflect the housing circumstances and domestic occupancy patterns of such households. Pensioners have higher-than-average energy needs as a result of spending much of the day at home, while being reluctant to undertake energy investment (see SSO, 2005; GRM, 2000).

However, the survey also showed that rather than flats, houses were associated with lower energy costs. This is demonstrated, for instance, by the unusually low cost-of-warmth figures for extended families in houses, and the lower number of such families living in 'insufficiently heated homes' in Isar (around 33 per cent of all households in the respective group, as opposed to ca. 53 per cent in Pripor). While it was not possible to ascertain the relative weight of district heating systems in this factor, there was some evidence that the presence of DH can reduce a household's energy costs and hardship: despite the similar numbers of single parents living in flats in both districts, the cost-of-warmth ratio for this group was significantly higher in Isar, which has no district heating (0.39 as opposed to 0.35 in Pripor). This situation provides a statistical confirmation of widespread anecdotal knowledge, that many households find themselves using a heating system that does not match their energy needs *vis-à-vis* the temporal and spatial occupancy of the home. Forum CSID's (2002) study on urban poverty in Macedonia concluded that many small towns in Macedonia are characterized by a high rate of electric storage heaters in multi-storey apartment blocks, which gives their tenants very few alternatives with regard to the choice of heating system, especially if the buildings don't have chimneys. A leading daily newspaper in Macedonia quoted a Skopje householder as follows:

> The onset of winter brings a five-fold increase in my electricity bills. We live in a two-bedroom apartment and we use an electric storage heater, which uses a lot of electricity, and thus we try to use it as rationally as we can. But if last year's cold winter repeats itself again, we will have to turn the heating to the maximum. The children are still small and we cannot save heat ... We live in an apartment block and due to its impracticality and the blocked chimneys, we cannot use fuelwood or oil for heating. But even the other options, such as floor or district heating, are not cheaper (Utrinski Vesnik, 2000).

To summarize, it appears that Macedonia has two categories of households with a disadvantaged position *vis-à-vis* energy poverty. The first of these would include the general group of families with low incomes: welfare beneficiaries, households headed by unemployed adults, multiple-children households, and families who depend on agriculture for all of their income. Among them, the first two groups are disproportionately urban, while the latter two tend to be concentrated in rural areas in the north of the country. However, a second group of vulnerable households is constituted by the families who are at risk by virtue of their housing circumstances, mainly pensioners and families with young children. In their case, the emergence of energy poverty can be attributed, in part, to the poor energy efficiency of the home, and above-average daily energy needs.

Although these findings cannot be verified though nationally-representative statistics – because of the absence of specialized energy poverty surveys – they nevertheless point to a common trend: energy poverty is contingent on a wide set of housing and social conditions, beyond low income. This makes it both a lower- and middle-class phenomenon in Macedonia. Such findings set doubt on Lovei et al.'s (2000) income-based classification of post-socialist populations into 'poor' and 'non-poor' groups, for the purpose of appraising the effectiveness and efficiency of different energy subsidy mechanisms.

Energy poverty patterns II: Czech Republic

Although both Macedonia and the Czech Republic cross-subsidized domestic energy prices during socialism, consequent price rises in the Czech Republic have been far less dramatic in relative terms. While residential electricity tariffs have been increased by more than 400 per cent since 1991, this figure represents only a 25 per cent rise above the inflation rate during the same period. At the same time, DH prices have increased by 340 per cent in absolute terns, over the 1991 level (CSO, 2005; ADHCR, 2006). Thus, DH and electricity have actually become cheaper in relation to the inflation rate, although the same cannot be said about natural gas (see Figure 5.4).

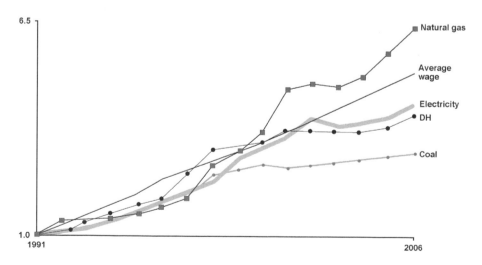

Figure 5.4   Czech Republic: Relative change in the average wage vs. various fuels used for domestic heating 1991–2006 (1991 = 1)

Source: ADHCR, 2007.

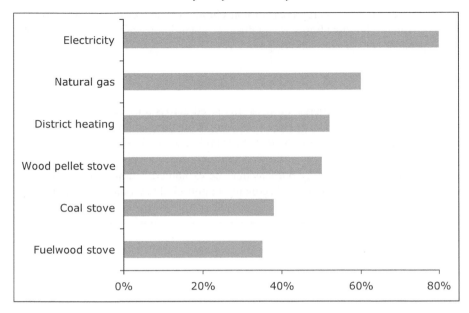

Figure 5.5   Czech Republic: share of energy costs dedicated to domestic heating in the cases of different fuels

Source: ADHCR, 2007.

The Czech Republic also has a more diverse fuel mix in the residential sector. Approximately 40 per cent of energy is provided by gas, with a remaining 50 per cent being split almost equally between electricity and heat (see Figure 5.1). According to the Association for District Heating of the Czech Republic (ADHCR), approximately 50 per cent of the population relies on DH for domestic warmth. However, there are major differences between rural and urban areas in this regard. DH and central heating are much more widespread in urban centres, while a quarter of rural households still use coal (AEA, 2006). This is largely due to the poor availability of DH and piped gas in villages, even though the latter is now the dominant source of heating in such regions, due to the rapid expansion of gas distribution systems in recent years.

At 85 per cent 'maximum possible' energy efficiency, gas-powered DH is considered the most efficient fuel, followed by coal-powered DH at 80 per cent, central heating at 78 per cent, individual gas heaters at 74 per cent, and electricity at 33 per cent (ADHCR, 2006). However, this is the maximum technical potential for energy efficiency, which is rarely achieved in practice: actual levels of energy efficiency are usually 5–10 per cent lower. The ADHCR furthermore claims that, in terms of monetary cost per units of final warmth, electric heating is the most expensive fuel, especially in apartments without storage heaters. It is about 30 per cent more expensive than DH and gas-fired central heating: The cost of warmth decreases by a further 10 per cent in the case of central heating systems that are fired with fuelwood (ADHCR, 2006).

The relatively high costs of electricity are reflected in the structure of household budgets. According to the ADHCR models, domestic heating accounts for more than

80 per cent of the total energy costs of households who use electricity as a primary heating source. The same figure is significantly lower in the case of natural gas, DH, and especially coal or fuelwood (Figure 5.5)

Energy poverty indicators

According to the HES, energy burdens increased across all Czech income deciles between 1995 and 2005. The greatest rises were in the third and fourth deciles, where the share of energy expenditure within the total household budget surpassed the 10 per cent mark in 2005. However, the first and second deciles also began to approach this level. Their lower energy burdens may be explained, among other factors, by the greater weight of the unofficial economy in household incomes (see Figure 5.6).

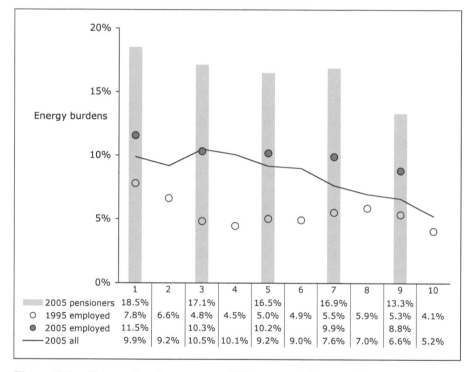

Figure 5.6   Energy burdens among different social groups in the Czech Republic

Notes: Author's calculations based on household expenditure data from CSO, 1996; 2006. 'Employed' stands for households headed by adults in full-time employment. 2005 data is available for quintiles only.

However, this general figure hides a more complex picture. For example, in 2005, the energy burdens of pensioner households were well above the 10 per cent mark in all five income quintiles, and reached nearly 20 per cent in the first quintile (see

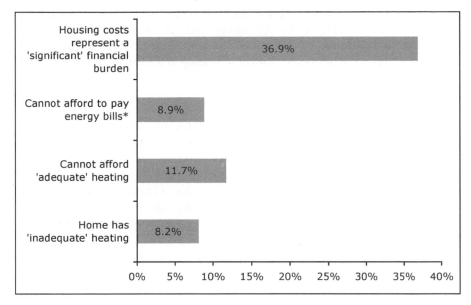

Figure 5.7   Subjective experiences of energy poverty, Czech Republic. Multiple answers were possible

Source: CSO, 2005.

Figure 5.6). Energy burdens were approximately 11 per cent even among households within the first quintile of families led by employed adults, who are normally better off than the remainder of the population. Ten years earlier, the same figure for the bottom two deciles within this group averaged 7 per cent.

In a paper dealing with the subject of energy affordability in the Czech Republic (Buzar, 2007a), I found that the compensating variation for the period 1995–2004 divides Czech households into three groups. While households in the bottom decile had decreased their energy expenditure by more than 5% in relative terms – a sure sign of energy poverty – those in deciles 2–6 had seen a relative rise in the total monetary sum dedicated for heating and lighting. While this increase is probably related to the rapid growth of energy prices since 1995, it also signifies that such households have allocated additional income for energy.

However, it also transpired that the energy expenditures of the top four deciles have actually fallen since 1995. This expenditure decrease is unlikely to have been associated with a drop in energy consumption, because the incomes of such households had increased at a greater rate than those in the lower deciles. Instead, it is more probable that the availability of cheaper and/or more efficient fuels, coupled with the improved technical quality of the residential stock, had helped reduce the energy bills of the richest 40 per cent of the population.

The concentration of energy poverty in the lowest income decile is verified by the surveys of well-being within the FBS, which have established that 8.2 per cent of households are not satisfied with the level of heating in their homes. This is slightly higher than the 4 per cent and 7 per cent of households estimated to be living under the APL and RPL, respectively (CSO, 2005). However, 37 per cent of households interviewed within the FBS well-being survey stated that housing costs represent a 'significant financial burden' on their family budget, while 11.7 per cent thought that they couldn't afford an 'adequate' amount of heating in the home' (Figure 5.7).

Socio-demographic profiles of energy poverty

Clearly, the Czech Republic has a more concentrated demographic structure of energy poverty, encompassing up to 10 per cent of the population. But the social, economic and spatial factors that are linked to the problem require further elucidation. In this section, I cross-reference a number of different sources that contain information about the presence of energy poverty-related difficulties within the Czech population. It has been impossible to provide a more direct demographic profile of domestic energy deprivation because a specialized study of energy poverty, simply, hasn't been undertaken to date.

Since domestic energy deprivation typically affects low-income social strata, it is useful to start from the general patterns of income poverty. One of the most helpful entry points into such an analysis is provided by Večerník's (2001) examination of the demographic makeup of households with incomes under the poverty line. It indicates that, in 1996, nearly half of the households whose total (or 'aggregate') income fell under the APL or RPL were represented by families with children. The same group composed more than 60 per cent of poor households in the equivalent-income classification.

The expansion of income poverty among the general population is confirmed by the recipient structure of the housing allowance, which was distributed to 333,500 families in 2001 (Personal communication, Czech Ministry for Labour and Social Affairs, 10 June 2002). Of the total number of recipient households, 33.3 per cent had five or more members, with one- and two-person households composing 40.6 per cent of the remaining beneficiaries. When placed within the context of the eligibility criteria of the allowance, this data suggests that multiple-children families (which are prevalent among the Roma minority) are disproportionately affected by income poverty, alongside single parent and lone pensioner households (Zoon, 2001; CSO, 2005).

The rate of change in the energy and housing 'burdens' (that is, the shares of energy or housing costs in the total household budget) can also help identify the groups vulnerable to domestic energy deprivation. The development of such trends during the post-socialist transition can be tracked through nationally-representative household expenditure data from the FBS. According to this dataset, the average housing burden of all Czech households increased from 9.2 to 15.4 per cent between 1995 and 2002 – an unprecedented rise in Western European terms, reflecting the growing gap between incomes and prices during this period of intense tariff reform (CSO, 2005). However, some social strata suffered ostensibly more than others. Hardest-hit were low-income young families and pensioners, as energy consumed more than 16 per cent of their net household income during 2002 (according to the methodological definitions of the FBS, all the households with total incomes equal to or less than 130 per cent of the social minimum are considered 'low-income families', see CSO, 2005).

The FBS also shows that there is a close correlation ($R^2=0.902$) between the shares of social assistance and energy expenditure in the total family budget. The share of energy expenditure is also weekly correlated ($R^2=0.366$) to the share of earnings from formal employment in household budgets (see Buzar, 2007a). The very existence of these connections, however strong or weak they may be, points to the broader significance of work-related income in the emergence of energy poverty. They indicate that all households with high energy burdens are less able to obtain their income from regular, formal employment, having relied on part-time economic activities, or the grey economy, to obtain the bulk of their domestic earnings. This situation has a specific geography, as, in general, welfare recipients in the Czech Republic tend to be over-represented in the regions of Northern Bohemia and Northern Moravia, and other regions with high unemployment rates (see Map 1.2). According to the FBS, low-income households are less reliant on district heating than average, as they tend to use fossil fuel-powered central heating and coal. In particular, low-income families with more than two children are the least likely to rely on district heat.

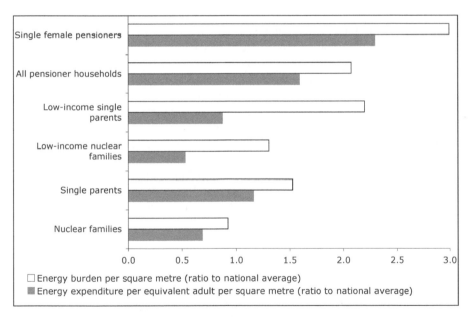

Figure 5.8   Energy expenditure per dwelling area for different Czech households in 2001

Source: CSO, 2002.

According to the FBS, pensioners are in a particularly difficult position with respect to energy price increases. The energy and housing burdens of pensioner households reached record levels in 2006. This means that they managed to retain constant levels of consumption of necessities, at the expense of non-essential services and goods. It has been suggested that this situation has arisen from prior experience with coping methods during periods of shortage (Večerník, 1995; Sik and Redmond, 2000). However, besides having the highest energy burdens, pensioners were also characterized by the highest energy expenditures per dwelling area, when adjusted for the total number of people using the space. Each square metre in a pensioner home attracted a three-times-above-average energy expenditure per person (Figure 5.8). Single female pensioner homes required the greatest energy expenditure in person-per-square-metre terms, although their aggregate energy expenditure per total apartment area was only negligibly smaller than average.

The ratios of energy expenditure per apartment area within the FBS also reveal the difficult domestic conditions of low-income families with children. Such households have been allocating 50 per cent less funds on energy per equivalent square metre than the national average, despite the increased burden on the family budget, exerted even by such low levels of consumption. Low-income single parents with children were in the most disadvantaged position, as the ratio of their energy burden to the aggregate dwelling area surpassed that of pensioners as a whole, although the equivalent energy expenditure per square metre was nearly 1.5 times lower.

Based on these trends, the question may be asked as to how energy expenditures and burdens vary according to total apartment size, especially since low-income and/or pensioner households in the rental sector live in flats that are much smaller than average (CSO, 2005). The only disaggregated dataset containing such information can be found within a study of housing expenditure (Obadalová and Vavrečková, 2000) based on the 1998 FBS, which included an additional 773 households with incomes at or below 130 per cent of the social minimum, alongside the basic FBS sample of 2515 households. This data reveals an anomalous situation, as it indicates that each square metre in the apartments within the 35–60 m² range was associated with a five-times-greater-than-average absolute expenditure per person, and a greater energy burden on the household budget (between five and two times higher than the largest flats in the sample, see Figure 5.9). It follows that families with energy affordability problems tend to be concentrated at the lower end of the apartment size range, although this conclusion should be interpreted with caution due to the small sample size of the survey.

The lack of representative quantitative data about thee socio-spatial characteristics of disadvantaged families makes it difficult to explain the reasons for the energy and housing expenditure patterns. Still, some deductions can be made on the basis of hypothetical models of the energy needs of different types of households, developed in 2000 by the Czech District Heating Association. Relating these models to 'real' FBS housing expenditure data has yielded surprising figures about the cost of the final energy service for different types of homes and families. It has emerged that families living in rented apartments spend the Czech Koruna (CZK) equivalent of 9.59 Euro per Gigajoule of domestic warmth, in contrast to the 7.45 Euro for households in detached family houses. Pensioners and nuclear households living in public rental housing (which is dominated by flats in tenement buildings) purchase energy services at the rates of 8.8 and 6.8 Euro/GJ, respectively. These findings contradict anecdotal knowledge, because they imply that the heating of smaller homes – particularly those in the regulated rental sector – requires a greater monetary expenditure per unit of delivered energy.

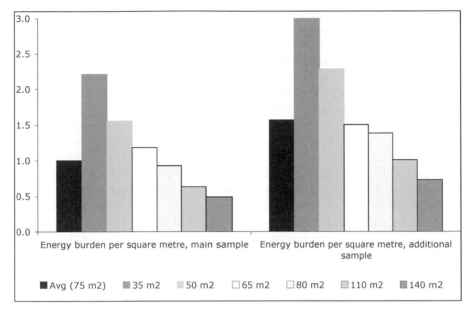

Figure 5.9   Energy expenditure per dwelling area in the households covered by the main and additional samples of the 1998 FBS, according to total apartment size (mean of main sample = 1)

Source: Obadalová and Vavrečkova, 2000.

Comparing these results to statistical surveys of housing costs shows that income-poor households, pensioners and urban rental tenants have been purchasing more expensive housing services than the remainder of the population. For instance, the housing costs for each square metre in the home of a rental household living in a large city were 32 per cent and 14 per cent higher than those of rental households in medium- and small-size settlements (Lux et al., 2003, p. 88). The monthly housing costs of owner-occupiers were 33 per cent, 34 per cent and 50 per cent lower than this amount, in large, medium and small settlements, respectively (ibid.).

Expenditure data from the Czech Statistical Office also point to the high energy costs of elderly people: the heating of each square metre in their homes has required a monetary expenditure of 1.13 Euro, more than 20 per cent above the average level for the whole country. It is telling that pensioner households (especially those consisting of single women) live in homes with significantly more floor space per each room. On average, the rooms of these households were 50 per cent above the Czech average, possibly, due to the fact that many of them occupy large, turn-of-the century inner-city apartments (CSO, 2005). This may have induced them to purchase less affordable energy services than other households, because a large room with whole-day heating (as such households have above-average domestic occupancy rates) is much more expensive in energy terms, compared to a smaller room in a flat that is not occupied during most of the day.

Last but not least, surveys of subjective well-being also emphasize that particular social strata are particularly vulnerable to poverty and deprivation, especially with respect to housing costs. For example, although 19.5 per cent of individuals over the age of 60 consider themselves 'poor' (CSO, 2005), Obadalová and Vavrečková's (2000) research indicated that 60 per cent of the respondents in this group felt that they cannot afford the basic necessities of life. Similarly, Lux et al.'s (2003) work found that the 'lack of adequate heating' is identified as a housing problem by 8.3 per cent of pensioners and 10.7 per cent of young families (households with a head of household under 35). However, the respective figures in the case of 'poor technical quality of the dwelling' were significantly higher – 15.1 per cent of pensioners and 20.7 per cent of young families (ibid.). The same survey established that a record share – almost 55 per cent – of the households living in urban housing estates or older tenement buildings feel that their housing burdens were 'very high', while the same figure reached 40 per cent in the case of households living in urban family homes (ibid., 89). This situation can be explained, in part, by the low construction standards and poor thermal insulation of the homes inhabited by low-income households, which implies that such families are facing disproportionately high running energy and housing costs.

In summary, then, it appears that the income dimension of energy poverty is stronger in the Czech Republic than in Macedonia, although housing infrastructures also play a role, especially in the case of pensioners. Nevertheless, households who are in a severe economically-disadvantaged position are also likely to be suffering from insufficiently warm homes. There is some evidence to suggest that energy poverty is over-represented among three distinct groups in the low-income band: single parents, multiple-children households, and pensioners.

Conclusion

The evidence reviewed in this chapter points to the existence of a direct link between energy poverty, on the one hand, and increasing energy prices and falling real incomes, on the other. Both Macedonia and the Czech Republic have experienced changes in household energy consumption and expenditure, following the abolishment of universal socialist-era subsidies of energy tariffs, and the drop in mean real incomes due to transition-related poverty and unemployment. Although some households have managed to move towards comparatively cheaper fuels, others have suffered from domestic energy deprivation.

But this is where the similarities end, as the socio-economic features of domestic energy poverty are strikingly different in the case of each country. In Macedonia, the population living in inadequately heated homes is clearly much bigger than the 30 per cent of households considered 'poor' according to the RPL set by the state, and may even include 60 per cent of all households in the country. Energy poverty in this country has thus assumed both a low- and middle-income character, because it extends beyond the boundaries of relative and absolute poverty lines. In other words, relative income poverty in Macedonia is a subset of energy poverty. Conversely, domestic energy deprivation in the Czech Republic is a socially-marginal phenomenon, because it exists between the APL and RPL. The rate of energy-poverty can range between 4 and 11 per cent of the population, depending on the way in which the problem is being defined and measured. The available statistical data suggests that several discrete socio-economic groups inhabit insufficiently warm homes. They are vulnerable by way of their specific socio-demographic circumstances. The total energy poverty headcount does not exceed 4.5 per cent.

Domestic energy deprivation in both countries has been triggered by a common predicament: the decreased affordability of energy in post-socialism. Households have responded to this situation in three ways. Some have continued to consume energy as before, because they have been able to afford it. Others have switched towards more efficient or cheaper fuels, because they have disposed of either the capital stock, or the necessary funds to make such a move. For a third group of households, however, most (or all) forms of energy have become unaffordable post-1990, even when they have been able to substitute fuels. These groups have been forced to decrease their energy purchases, in some cases below the biologically-acceptable limit, which means that they have been pushed into energy poverty.

Moreover, in Macedonia and the Czech Republic alike, energy poverty affects all the typical socially marginalized strata: low-income families headed by unemployed adults, households with irregular sources of income, single parents and multiple-children families. In all these cases, domestic energy deprivation is a function of low equivalent incomes, which stem from a wider array of social-exclusion related processes, the analysis of which would extend beyond the aims of this book. A second layer of vulnerability, however, is constituted by the households with disproportionately high energy needs, such as families with pensioners and children. Their demographic circumstances may increase their energy expenditure and burdens beyond tolerable levels.

Relating these findings to the conclusions reached by Chapter 3 implies that the conditions found in Macedonia and Czech Republic could be replicated elsewhere, in countries where, respectively, a 'pervasive' or 'insular' geography of energy poverty factors is present. In relation to the outcomes of Chapter 4, however, they indicate that the divergent nature of the two countries' institutional systems has produced different patterns of energy poverty. Explaining this dissimilarity through an evolutionary economic approach – in particular, the theory of non-convergence of economic systems due to high transformation costs, outlined in Chapter 1 – would imply that the extent of energy poverty in the Czech Republic has been limited (in generic terms) by the systemic capital of institutional modes inherited from the pre-socialist period, when the country was a market-based democracy. In other words, the antecedent elements of Western capitalism managed to co-exist with socialist institutions, facilitating the system's adaptation towards a market economy during the transition, and easing the energy poverty burden.

The same theory would connect Macedonia's extensive energy poverty problem with the lack of Western-style institutions in this country prior to the post-socialist transformation. Even though Yugoslav socialism was of the most 'marketized' type, the imposition of a neoliberal regulation to an underlying socialist model resulted in systemic disruption due to institutional resilience and non-convergence, and economic shocks. Energy poverty is a manifestation of these discrepancies.

The divergent institutional development of the two countries can also be interpreted with the aid of a political economy-based approach. Such an analysis would focus on the link between the creation of a formal institutional infrastructure, on the one hand, and the day-to-day decision-making process, on the other. It implies that the superficial transplantation of a given regulatory solution on an institutional environment resistant to market reforms can hinder their effective implementation, creating inefficient organizations and relations that may promote energy poverty. The expansion of energy poverty in the Czech Republic may have been hampered by the existence of stronger institutions, lesser external and internal constraints, and a broader societal consensus about the necessity of market-based reforms.

However, although such theories can help unravel the complex institutional roots of energy poverty, they do not fully account for the spatial and social relations that have rendered some social groups more vulnerable than others. In the following chapter, I look at the lived experiences of energy poverty, with the aim of exploring the micro-scale circumstances that drive domestic energy deprivation.

Chapter 6

Everyday Experiences of Inadequate Warmth in the Home

So far, this book has addressed questions relating to the policy background of energy poverty, and the broader socio-economic reasons for the observed disparities between Macedonia and the Czech Republic. But institutional legacies and cultures *per se* are insufficient to explain the markedly different size and structure of the demographic groups affected by energy poverty. As a result, the principal aim of this chapter is to delve into the 'grain' of social relations, by examining the everyday practices and articulations of energy consumption at the level of the household. I focus on the micro-scale experiences of domestic energy deprivation in Macedonia and the Czech Republic, by connecting the policy mechanisms found in the previous three chapters (strategic selectivity, institutional trap, energy efficiency gap) with the everyday lives of the households affected by energy poverty. Special attention is paid to the interaction of fixed and mobile infrastructures in the residential energy sector, and the manner which the lived experiences of energy poverty are shaped by domestic occupancy and energy efficiency patterns.

The chapter is based on ethnographic evidence gathered during on-site research in Macedonia and the Czech Republic. A total of 20 interviews (10 per country), each lasting between two and four hours, were carried out between December 2003 and January 2006. In Macedonia, the interviews were concentrated in the districts of Pripor and Isar, which I described in the previous chapter. As for the Czech Republic, I interviewed different range of households in different parts of Prague, in addition to Chomutov – a town of 52,000 inhabitants in North Bohemia (see Map 1.2). Although Prague is one of the most prosperous and rapidly growing urban agglomerations of Central Europe, it contains significant economic disparities and layers of deprivation. Chomutov has been facing severe economic problems due to massive unemployment, caused by the closure of large industrial enterprises in the early 1990s and the lack of consecutive investment in the tertiary sector.

The interviews were of a semi-structured character, covering issues such as: the housing and social biographies of the respondents, their residential housing situation, social ties, daily mobility patterns, consumption requirements, economic status, sources of income, involvement in community life, and subjective attitudes towards spatial, social and political changes in the urban environment. Access to the interviewees' homes was gained through informal personal networks. The interviews were tape-recorded with permission and later transcribed. Although I attempted to include representatives of the widest possible range of demographic and housing situations in the interview list, it was impossible to match the choice of interviewees with the household structures of the study areas, because of the lack of statistical data at that scale. Thus, rather than providing anything like a wider socio-economic analysis, the chapter is aimed at highlighting some of the generic social and spatial issues faced by such households in the negotiation of their everyday lives.

Voices of the energy poor I: Macedonia

It was difficult to decide which types of households to interview in this country, due to the pervasive nature of energy poverty. The final list of interviews included a wide range of experiences and situations, as each of the surveyed households had a different family structure, housing situation and social status. Nevertheless, some common features can be identified. For example, all of the interviewees were deeply dissatisfied with the services and policies of energy companies. It was constantly pointed out that electricity prices were too high in comparison to family earnings, and that 'tariffs should be adjusted to reflect the living standard of the Macedonian population' (Personal communication, 22 December 2004). Also, 'why should we be paying Western prices, while not receiving Western wages?' (ibid.).

The interviewed families in Skopje also shared a second dissatisfaction, not with the price, but rather the quality of district heating services:

> Last month, the temperature in our flat ranged between 18 and 19°C during the day, and 15°C at night. Although there are radiators in each room, they were lukewarm and we had to use electric heating. I called *Toplifikacija* to find out why this was happening. The person at the other end of the line told me that they can guarantee a temperature of 19°C. I argued with him and said that they are legally obliged to heat the homes up to a temperature of at least 20°C ... I had read that in the newspaper the previous day. He would not yield. A while later I called his boss, who confirmed that I was right and promised to despatch a team to our building to find out the source of the problem (Personal communication, 13 January 2006)

It also transpired, somewhat unexpectedly, that even though the interviewed households were unhappy about the price and quality of energy services, none of them perceived the right of access to affordable warmth as a positive entitlement: 'Of course we should all have the right to warm homes, but the energy we get has to be produced somewhere, right? Someone has to pay the bill for that. I just wish that we did not have to pay the entire bill' (Personal communication, 21 December 2004).

The predominance of such opinions in the qualitative sample may reflect a deeper cultural trait, in that Macedonians have very low expectations from utility monopolies, due to receiving poor quality services for decades under socialism. However, it should also be taken into consideration that this situation could have arisen simply as a result of the composition of the interview group.

The surveyed households had adapted to their difficult domestic energy circumstances in a wide variety of ways. One of the most surprising findings was the realization that most interviewees possessed a sound basic knowledge of the flow of energy in their homes and the technical methods for its management. An oft-used coping strategy, therefore, was the application of low-cost energy conservation measures to minimize the loss of delivered warmth in household appliances and through the building fabric: 'We are constantly trying to save as much energy as we can. We use only 60W light bulbs and the storage heater is off whenever possible. Also, we have been adding some cheap draught-proofing to the windows on the north side of our apartment' (Personal communication, 20 December 2003)

Among the interviewed households, pensioners had the most comprehensive knowledge of energy conservation measures, including: passive solar heating, extensive use of curtains and carpets for insulation, as well as the simultaneous reliance on fuelwood stoves for heating, cooking and hot water preparation. Often, the chimneys were run inside the dwelling for as long as possible, to increase the circulation of redundant heat. Yet the World Bank identified less conventional coping methods in Macedonia:

> Families try to conserve by using the wood only to cook a midday meal, often spending the rest of the day without heating. People experience having their electricity cut as shaming, and if they have no money at all, try to borrow from friends or relatives only to avoid being cut. Some households simply live without electricity. A young couple living in Sveti Nikole, near the city of Štip, live in an old house on the outskirts of town, where they don't receive electricity. Instead, they provide themselves with weak lighting by using a small car battery in one of the rooms (World Bank, 1999c:124).

The most widespread response to energy non-affordability was the reduction of domestic energy consumption. In all of the interviewed households (except those relying on district heat) entire-day heating was limited to the few rooms where most of the daily activities took place. Moreover, the vast majority of interviewees (more than 80 per cent) in the Štip and Pripor surveys heated less than half of their homes in winter. This is similar to Kovačević's (2004) findings for Serbia and may reflect a broader regional trend, which implies that the colder period of the year induces a two- to three-fold decrease in the usable dwelling area of the housing stock. Such 'hidden' spatial contractions have major conceptual and policy implications, considering that 'space in the home is of little value unless it can be used as a living-space' (Boardman, 1991, p. 12).

Still, there are instances where the reduction in the size of heated space was insufficient to overcome the affordability problem, leading to reductions in the level of heating *per se*. Among many of the interviewed households, average indoor temperatures ranged between 17.5 and 19°C at the time of the interviews. The local press is replete with more drastic examples, even among the inhabitants of state-supported pensioners' homes:

> I use an electric storage heater. When it is cold outside, the heater must be on almost during the whole day. My pension is 4,500 denars, and the electricity bill is 2,000 denars per month. Last month's bill was 4,000 denars. Those pensioners with 6,000 denars can somehow get by, but it is very difficult for us in the lower band. What am I supposed to do now? I often starve in order to pay my energy bills. If I had district heating, I would have money for food and I will be warm all day (elderly female pensioner, Dnevnik, 2002)

An extended family in a rural detached family house

The experiences of households living in inadequately insulated homes can provide valuable insights into the relationships between energy poverty and residential energy efficiency. In Pripor, one of the most illustrative case studies was offered by the 'Stankovi' household, an extended family consisting of two pensioners, a married couple of unemployed adults, and two children (15 and 6 years old). At the time of the interview, they were living in a two-story brick house at the edge of the city, in a location that allowed them to supplement their earnings through subsistence agriculture. The only regular monetary income was in the form of one pension and a limited amount of social assistance for the two unemployed adults. This was one of the main reasons why they were unable to pay their electricity bills on time, as the electricity company would only accept cash payments.

Another causal factor of this family's energy affordability difficulties resided in the low efficiency of its domestic appliances, most of which had not been replaced since the 1970s. The fact that nearly all householders spent most of the day at home meant that the house had to be heated during the entire day. The careful use of energy couldn't offset the loss of heat through household appliances and the building fabric. As a result, winter-time temperatures in the upper floor of the house rarely reached 18°C. This forced the family to spend most winter mornings and afternoons in the living and dining rooms on the ground floor, which were equipped with fuelwood stoves. In the evening, the pensioner would move to the upper floor, which would already be warm due to the convection of warmth from the ground floor, and the use of open metallic chimneys which radiated heat from the stoves below.

The predicament of this household illustrates the role of housing and energy infrastructures in the production of energy deprivation and poverty. Had they lost less heat through the building fabric and domestic appliances – or spent less of the day at home – the 'Stankovis' would have been considerably less vulnerable to energy poverty. Even minor improvements of the thermal insulation of external walls and windows could have reduced this household's energy costs by a third, allowing them to spend the extra income on, for instance, heating the upper floor of the house, or burning the fuelwood stoves for a longer time.

Elderly couple in a suburban detached family house

The domestic circumstances of the 'Selmani' family exemplifies the socio-spatial embeddedness of energy poverty. Their meagre monetary income was insufficient to cover the disproportionately high heating bills, which stemmed from the loss of useful warmth through the windows and walls, as well as the ageing fuelwood stoves and electric heaters. Moreover, the two pensioners required a higher-than-average amount of heating, because they spent a long time at home in the winter – they noted that 'there are practically no indoor public spaces in our neighbourhood and we cannot afford to travel to the centre of the city'. They were forced to reduce their heating consumption by approximately a third below what they would normally need, and were facing health problems as a result:

We mostly rely on the electric heater. When it is cold outside, the heating must be on almost during the whole day. But the high energy bills force us to wear coats, stay in bed all day wrapped in blankets, or to visit friends who can afford to use more energy (Personal communication, 21 December 2004)

The disproportionately high (30 per cent) energy burden of this pensioner household can be attributed to its domestic circumstances. Two people lived in an area of 90 m², although they could barely afford to heat two rooms in winter. They spent most of the day in the north-facing kitchen/dining room area on the ground floor, which was equipped with both an electric and a fuelwood stove (see Figure 6.1). At night they moved to the southeastern bedroom, which had a weak electric heater. The fuelwood stove in the large living room was operated only when guests were present. The other bedroom was mainly used for storage purposes, as it was cold and unoccupied during at least half of the year.

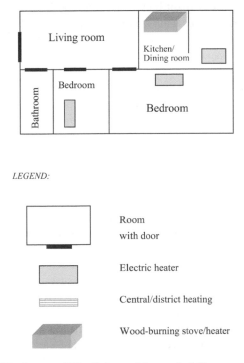

Figure 6.1   The home of the Selmani household*

*Figures 6.1–6.5 are for illustrative purposes only. They are based on visual observations of the spatial structure of the surveyed homes, rather than precise measurements.

This household also tried to minimize the use of electric appliances: they even heated the bathing water on the wood stove in the kitchen. Yet even with such high levels of energy conservation they consumed a considerable amount of energy, due to their house's poor wall insulation. Most of the generated warmth was lost through the outer walls and internal partitions, as evidenced by the fact that temperature in the main dining room at the time of the interview was 19°C (at an outside temperature of -1°C) although the stove was at full power. The pensioners noted that this was the typical room temperature during the day: 'perhaps it could be a bit warmer, but we wear warm clothes'. However, they did point out that the evening temperature in the bedroom was 'uncomfortably cold', as the electric heater could not generate warmth at a faster rate than was lost from the house. They complained of respiratory and blood circulation problems as a result.

The interviewees were aware that the low thermal insulation of the walls was the main reason for the disproportionate cost of warmth, but they were reluctant to do anything about it: 'We have made some improvements to the draught-proofing of the northerly windows, but it would be too expensive to add an additional layer of insulation to the walls. Besides, we would not feel comfortable to let builders inside and around our house' (Personal communication, 21 December 2004).

Each month, this household spent between 25 and 60 Euro (the amount varied seasonally) for heating their home, against a monthly pension of 140 Euro. Although they could have saved up to 20 Euro per month with the aid of basic energy efficiency measures, the necessary disposable income was clearly lacking, and they could not get any support from either the state or private banks for such purposes. The predicament of this family embodies the vicious circle that feeds energy poverty, because the high running costs for domestic heating deprived them of the monetary means for making the investment that would reduce the running costs in the first place.

The experiences of these households are related to a wider problem at the level of the entire neighbourhood. Similar to other major towns that experienced a rapid urbanization during the 1960s and 1970s, many houses in the districts of Pripor were built as extensions to older family homes, in an attempt to minimize construction costs. This housing is of smaller size, poorer quality and lower energy efficiency than average, because the desire to build as quickly and cheaply as possible led many households to omit standard energy-saving measures, such as double-glazing and cavity wall insulation (SSOM, 2001). But the reduction in the initial capital cost has been offset by a penalty on running energy expenses, as the cost-of-warmth ratio among extended families living in detached or terraced houses is significantly lower than the analogous expenditure among the same type of households in collective apartment buildings. Thus, the poor construction quality of these homes makes them thermally- and economically-wasteful, and pushes their inhabitants into deprivation.

Extended family in an inner-city detached family house

The 'Petkovi' household faced similar problems. This was an extended family of one working adult, two school-age children, and an elderly couple, one of whom received a pension. Although they lived in a large detached family house of 150 m², they could only afford to heat only one room in winter. They were purchasing energy at a very high cost, due to the extremely poor thermal efficiency of the building fabric, and the reliance on outdated, energy-wasting electric storage heaters. The built structure of the house did not allow for the installation of a fuelwood stove, or any other sources of energy that would be cheaper than electricity:

> We haven't paid our bills for five years. We asked the electric company to 'forgive' the debt ... Instead they offered to divide it in 10 payments. I keep begging them not to turn the electricity off ... We can barely afford the electric storage heater for the small room where we spend most of the day (Personal communication, 20 December 2004).

During the interview, the 'Petkovis' expressed concerns about the thermal insulation of their home. They were particularly unhappy about the windows and the roof which, they thought, was in need of urgent repair. There were also above-average heat losses through the outer walls. The cost of warmth was especially driven up by the inability to switch away from electricity in the two southern rooms with new capital investment. But the interviewees were not certain as to how they should approach this issue, stressing that they were looking for energy efficiency advice.

Thus, the mismatch between 'rigid' heating infrastructures and 'fluid' housing needs forced these two households to pay a higher price for each unit of warmth delivered to their homes. This is one of the main reasons why they were suffering from inadequate levels of domestic heating. The relationship between poverty, housing and energy costs implies that the embeddedness of particular heating systems within particular housing infrastructures can act as a poverty-inducing factor and a socio-technical constraint (compare with Law and Bijker, 1992). The structure of the heating system *per se* prevent households from switching to more flexible, affordable and efficient sources of energy. To use the language of actor-network theory, this relationship allows a particular network of non-human actants to exercise agency over the household scale.

Nuclear family in an inner-city apartment building

On-site interviews pointed to another mechanism through which housing infrastructures interacted with household energy consumption to aggravate the circumstances of vulnerable households. It emerged that energy poverty can arise as a result of the mismatch between heating systems and occupancy patterns. This was particularly pronounced in the Isar area, which contains a disproportionately high number of households relying on electric storage heating; they represented 45 per cent of the total number of interviewees, against a national average of approximately 25% (AEA, 2006). Some of the most interesting findings in this district came from individual examples of the leading two groups suffering from insufficiently warm homes: nuclear families with unemployed adults and/or young children, and extended families living in detached or terraced houses.

The 'Mihajlovi' family can be classified into the former category, as it contained two adults (one of whom was unemployed), and two children under the schooling age of 7. At the time of the interview, they lived in a small apartment on the fifth floor of a tenement building in the centre of Štip, and relied solely on electric power for heating and lighting (see Figure 6.2). Their income was below average for Macedonian standards, and they were unable to make major capital investments in the apartment that they owned. The fact that the home was unoccupied during the day – when the children attended kindergarten and the parents were at work – had an ameliorating effect on the energy burden. Yet this was insufficient to offset the cost penalty of the electric storage heater, which drew power at night and had to be kept on constantly during the day, in order to obtain heat in the evening, when the children needed it most. As explained by the head of the household:

> We rely exclusively on electricity for heating and lighting, and our costs are considerable ... It would be nice if our block had chimneys so that we could use wood for heating, because the storage heater is more and more expensive. It must be turned on during the entire day, although we are at home only in the evenings. Also, it diffuses the warmth throughout the flat, and we only need to heat one room' (Personal communication, 18 December 2004).

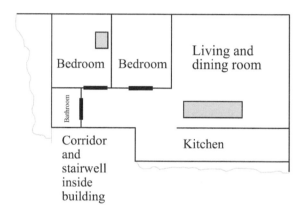

Figure 6.2   The home of the Mihajlovi household

Voices of the energy poor II: Czech Republic

The specificities of energy poverty patterns in this country warranted a different household selection and interviewing process, including a heightened focus on pensioners and extended-family households. The ethnographic interviews revealed many unexpected differences and similarities among various scales and contexts. Similar to their Macedonian counterparts, Czech pensioners also reported health problems, while complaining about the low affordability of energy. However, none of the interviewed elderly persons questioned the services, prices and disconnection policies of utility companies. Instead, great value was placed on paying energy bills in a prompt manner: 'it is shameful when one is not able to pay electricity and heating bills on time' (Personal communication, 20th December 2005). Such statements echo the findings of similar poverty assessments in Central Europe, which found that energy debts and arrears were associated with a negative social stigma (UNDP, 2003; Sik and Redmond, 2000).

Many of the interviewees reported major energy efficiency problems associated with low levels of investment, such as: outdated household appliances, draughty windows and doors, and rusty and leaky DH pipes and radiators. Although these families were aware of the loss of thermal energy due to the chronic lack of housing maintenance, their responses to this situation were different. The single mother whose apartment is in a Prague inner-city block didn't feel that she could address the situation: 'I can tolerate this as long as I can survive somehow … I suppose there is nothing I can do without the municipality's help' (Personal communication, 21 December 2005).

A similar attitude was expressed by the elderly couple in municipal rental housing in Chomutov, although they did point out that the tenants in their block had recently agreed to ask the municipality for financial support towards renovating the heating and construction infrastructures. The second pensioner household – a single female pensioner living in a private house in Prague – simply stated that she has 'hardly enough money to survive, let alone maintain the heating system'. As a coping strategy, she had decided to heat only one room and reduce the indoor temperature to 17°C during the day. Still, despite these differences, both pensioner households were linked by a common opinion: that the limited nature of their remaining lifespan discouraged housing investment. The lack of ownership rights supplemented and reinforced this underlying attitude.

The domestic circumstances and attitudes of the two Roma households were strikingly different. Besides living in extremely substandard municipal housing, the households derived all of their income from social aid and black market activities, which placed them in a very difficult financial situation. Housing investment or residential mobility was not on their list of priorities. Similar to some of the Macedonian interviewees, there was a distinct dissatisfaction with social welfare institutions. In general, the interviews echoed Zoon's (2001) findings about the housing conditions of the Roma minority in the Czech Republic.

Pensioners in an inner-city apartment building

'Jan and Petra Víšek' were a retired elderly couple living in a municipally-owned prefabricated block in the depressed 'rust-belt' town of Chomutov (see Figure 1.1). Having worked in chemical and coal mining plants most of their lives, both of them now received regular pensions. Thus, unlike most of the inhabitants of the area (the unemployment rate in Chomutov is above 20 per cent) they had a steady source of income, which allowed them to pay their energy bills on time (they felt that this was extremely important). However, this household did not have a sufficient income to cover their energy needs, which were disproportionately large due to the high occupancy rate and low efficiency of the building. It was easily evident that they had been unable to make major capital investment in the home. For instance, the electric and gas appliances had not been replaced for more than 10 years. This was one of the main reasons for the 'Víšeks'' excessive electricity costs. They also tried to minimize their use of district heating and gas (which was used for hot water in the bath), for fear of running into energy debts.

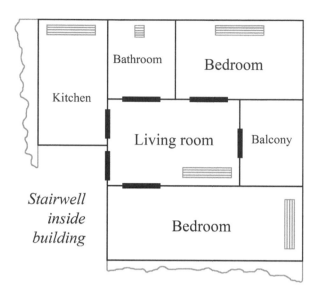

Figure 6.3   The home of the Víšek household

In addition, the interviewees complained that the district heating was often unable to compensate for the heat loss through the poorly-insulated walls, particularly in the northeastern bedroom which was at the corner of the building. As a result, they had decided not to heat that room, choosing to spend their time in the kitchen, living room and southern bedroom instead (see Figure 6.3).

Extended family in a suburban apartment building

The 'Sabek' family consisted of one pensioner, a married couple with one unemployed adult, and three dependent children. Thus, they had above-average energy needs due to their specific housing situation. This family lived in a municipally-owned building inhabited mainly by low-income groups. Although the flat was relatively well-kept, the apartment block was substandard and dilapidated. The plaster on the external walls was falling off, revealing the lack of insulation: 'The entire building is in a very poor state. We try to keep our apartments in good condition, but nobody has invested in the refurbishment of shared spaces – staircases, the façade, roof ...' (Personal communication, 17 December 2005).

The consequences of the substandard quality of the construction work – especially the lack of external wall insulation – were evident in the two south-facing rooms, which had slightly lower temperatures at the time of the interview, despite the presence of central heating (see Figure 6.4). This, together with the family's specific demographic situation, carried the principal blame for the 'Sabek's' exorbitant energy bills. The interviewees faced constant energy and rent debts, and feared that they might be evicted even from this flat, in which case they would have 'nowhere to go'.

Corridor and stairwell inside building

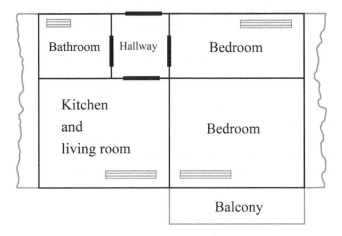

Figure 6.4   The home of the Sabek household

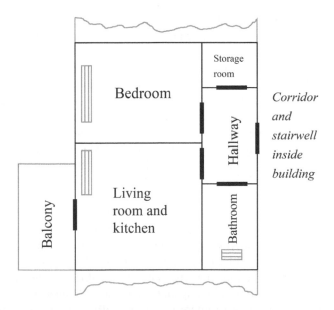

Figure 6.5   The home of the Korynta household

Single mother in an inner-city apartment building

Single parents emerged as one of the groups most vulnerable to domestic energy deprivation. This may be attributed to a combination of above-average housing needs, poor housing quality, and particularly, low income. The situation of 'Ms. Tereza Korynta' (Figure 6.5) illustrates the problems faced by such households.

'Tereza', the head of the family, lived in a newly-built private rental flat in the gentrifying district of Žižkov, just to the east of the city centre. She worked in a small private company every day between 9 and 5, during which time her 7-year old son stayed in a nursing home. Although her salary did cover the rent, she 'never managed' to pay the district heating and electricity bills on time. This was partly a function of her increased energy costs, which stemmed from the poor efficiency of the construction work: 'The flat was renovated before I came in, there was new flooring and the windows had been retrofitted, but it's no use, this is an old building with no insulation in the walls' (Personal communication, 16 December 2005).

The low efficiency of the walls was the main reason for the high price of warmth, as she had moved into the flat recently and the electric appliances were new. The only solution for Ms. Korynta would be to increase her income. But she did not qualify for social aid, as her monthly earnings were just above the threshold set by the state.

Conclusion

The everyday experiences of families and individuals living in inadequately heated homes demonstrate the importance of household-level functionings in shaping domestic energy deprivation. Thus, the application of Sen's (1980) theory to the empirical material reviewed in this chapter would connect the decreased energy utility of energy poor households to the loss of socio-spatial capabilities, that is, the inability to use personal knowledge, skills and fixed housing capital as a means of maintaining well-being. However, it is also evident that domestic energy deprivation is linked to economic hardship. The lack of disposable income has a dual effect, because it limits the ability to purchase the needed amount of energy in the short term, while impeding long term investment in housing and energy infrastructure. The interviewed households expressed a deep dissatisfaction towards the central state in both countries, as well as municipal governments (Czech Republic) and energy companies (Macedonia), which were believed to carry the principal blame for their difficult situation.

The reviewed evidence has also raised a number of additional questions. It remains unclear to what extent energy poverty is produced by the lack of energy efficiency information, as many energy poor households appear to have a sound basic knowledge of insulation and heat flows. A second key dimension of is the role of personal initiative in overcoming poverty, that is, where does the cognitive threshold for action lie? A common thread linking all the interviewed households in the Czech Republic is the apparent lack of interest in energy efficiency improvements. This has transpired despite the households' awareness of the problem and the ability to procure financing from third parties (such as municipal and central governments). The point at which awareness translates into action (a frequent discussion topic in the energy efficiency literature – see, for example, Jaffe and Stavins, 1994), as well as the culture- and region-specific aspects of such decisions, remain open for discussion.

These findings build on the conclusions reached in Chapters 4 and 5. Although it was pointed out that insufficiently warm homes are associated with all the typical demographic groups suffering from social exclusion – households with unemployed adults, women, as well as ethnic and age minorities – it is also true that the manner in which heating systems, appliances and dwellings are used by consumers also has an effect on energy poverty.

There is a permanent contradiction between the rigid disposition of housing infrastructures and the ever-changing character of families' and individuals' domestic energy requirements. For example, the cost of domestic warmth may be increased by the mismatch between energy infrastructures and occupancy needs. In this case, domestic energy deprivation stems from the agency of heating systems as social constraints. The hardship of disadvantaged households arises from the inability to switch away from unaffordable or inflexible heating system, as result of, to a large extent, the unbalanced provision of housing and heating infrastructures during the socialist modernization of the 1960s and 1970s, and the rise of poverty and inequality during post-socialism. The collapse of socialist development ideals has 'trapped' vulnerable families in poverty-inducing domestic arrangements, whereby the home itself acts as a causal factor of deprivation. This is illustrated by the situation of Macedonian households living in tenement blocks which have no space heating option other than electric storage, although this is one of the most expensive modes, and is often not suitable for families which do not require whole-day heating. Czech single parents also face a similar predicament.

Thus, in addition to being income poor, disadvantaged households are also often 'flexibility poor', due to being unable to accommodate the rigid built and energy infrastructures of their homes to the fluid energy needs of different family members. But even if the configuration of the dwelling is ideally suited to a household's energy requirements, it may lack information about the flow of energy in the home. Poor knowledge about residential energy efficiency can result in the wasteful use of energy, which in turn increases the cost of domestic warmth. It is important to note that both the occupancy and information dimensions vary across space, because different household have different energy needs and knowledges.

In summary, the evidence reviewed in this chapter suggests that the institutional production of energy poverty is reinforced by the geographies of residential energy use among disadvantaged households. This conclusion has at least two implications: First, it means that dividing the population into discrete categories, based on aggregate or equivalent household income, might underestimate the size of the population living in the less extreme forms of energy poverty. Second, the linkages between domestic energy deprivation, on the one hand, and the spatial extent of social, economic and physical infrastructures, on the other, suggest that this is both a path-dependent and path-shaping phenomenon. Not only do antecedent policies and patterns of spatial organization influence the manner in which energy poverty is produced, but it is also true that the practices and lifestyles of energy poor households also affect the utilization and development of space per se. This is an additional institutional 'loop' of energy poverty, functioning parallel to the processes established in Chapter 5.

Chapter 7

Linking Conceptual Threads, Looking Towards the Future

In its entirety, the evidence presented in this book points to a mounting 'hidden' geography of energy poverty in the former socialist countries of Eastern and Central Europe. Although the problem varies drastically across different spaces and contexts, it clearly does not conform to the conventional definition of poverty. Instead, domestic energy deprivation arises via an interaction of inadequate incomes, poor housing, and specific household circumstances. Its causal factors are deeply networked and embedded in the economy, which implies that the emergence of this condition has far-ranging implications for post-socialist energy, social welfare and housing reforms, among other issues. It can be argued that many of the main structural problems in the energy sectors of transformation states are linked to energy efficiency, poverty and affordability.

Yet such conceptual connections are largely absent from both the scientific literature and policy praxis. This final chapter of the book outlines some of the key theoretical and decision-making challenges raised by the reviewed data. Before outlining the possible paths of future research and policy work, however, I will briefly reiterate the key findings made by the previous six chapters.

Summary of principal findings

In Chapter 2, I argued that the emergence of post-socialist energy poverty is rooted in the antecedent deficiency of theoretical knowledge and empirical experiences. The failure to perceive all energy affordability-related problems under the common heading of 'energy poverty' has prevented scientists and policy makers from seeing its causes and consequences in an integrated manner. Moreover, social protection mechanisms in the post-socialist states are assessed via a reductionist view of poverty, and there is a gap in the literature with respect to the role of international and local institutions in producing inequality. Even though the late 1990s have seen the publication of several sizeable studies dealing with the environmental and economic dimensions of inefficient energy use, much remains to be done in terms of incorporating this argument into the mainstream discourse on housing and social welfare transformation.

Chapter 3 examined, within the limits of data availability, the spatial and social patterns of poverty, health, energy efficiency, energy affordability, and the outcomes of energy sector transformation in different post-socialist countries. It emerged that the states which have been reluctant to reform their energy and social sectors possess a 'potential' geography of energy poverty factors. In these spatial contexts, household energy is still heavily subsidized, and the energy sector does not operate according to full market principles. Although many utilities in such countries face major bill recovery problems, these are not necessarily related to affordability. However, forthcoming energy restructuring efforts in this region may push significant numbers of households into energy poverty. The remaining ECE states – where energy prices have been rebalanced – contain either 'insular' or 'pervasive' distributions of energy poverty factors, depending on the strength of the social safety net, and the speed of general economic recovery.

The remainder of the book was dedicated to a comparative case study between Macedonia and the Czech Republic. In Chapter 4, I established that the emergence of energy poverty in these two countries is linked to the operations of governmental, civil and private institutions. Domestic energy deprivation is contingent on self-reinforcing political and economic mechanisms, such as the strategic selectivity of the state, the institutional trap, and the energy efficiency gap. The two countries' differing patterns and structures of energy poverty may be linked to their external reform environments, as well as the internal path-dependencies of their economic systems.

Chapter 5 illustrated these broad conclusions with more detailed statistical evidence from Macedonia and the Czech Republic. I found that more than 50 per cent of Macedonian households may be suffering from domestic energy deprivation, while the same figure hardly reaches 5 per cent in the Czech case. This means that dividing the population into discrete categories based on aggregate or equivalent household income may seriously underestimate the number of people living in the less extreme forms of energy poverty. The socio-demographic profiles of households with poorly heated homes indicate that energy poor households live in low-quality and inadequately maintained housing, while many of them have a specific domestic and family situation.

Chapter 6 examined the patterns of residential energy use among disadvantaged households. It transpired that low-income households in investment-starved homes, and families living in dwellings that do not suit their housing and occupancy needs, spend an above-average amount of monetary income per unit of energy. They are suffering from a mismatch between their energy requirements, on the one hand, and the residential infrastructures that they occupy, on the other.

Key trends emerging from the analysis

In analysing the extent of energy poverty across different contexts and spaces, I have aimed to develop a theoretical framework for establishing the social, economic, political and economic implications of the problem. This conceptual matrix consists of three layers, which emphasize the connections between structural economic and spatial legacies, on the one hand, and organizational cultures and decisions, on the other. Its first stage involves a background review of the energy, housing and social welfare reform process. This serves as a basis for the second component, which involves an analysis of the size and structure of energy poverty based on data about household expenditure, energy bill non-payment problems, as well as the households' subjective perception of the level of achieved warmth in their homes. The outcomes of these empirical analyses are connected to third component, which deals with the driving forces of energy poverty, as well as the institutional and socio-spatial conditions at different scales.

This conceptual framework has established a connection between reform obstacles in the energy sector and the causes and consequences of domestic energy deprivation. In the 'lagging' transition states, the national governments' fear of the social repercussions of radical energy reforms has created an under-reform trap, which is one of the main reasons why utilities are facing technical and financial problems due to non-payment and low prices. This condition in turn leads to energy crises and massive service cuts during periods of unusually cold weather. Conversely, the 'reformist' transition countries, which have been more forthcoming in implementing tariff increases and energy restructuring – albeit each has chosen its own path and method of transformation – are likely to face greater energy affordability problems if they have not developed an adequate social safety net. In both situations, the existence of insufficiently warm homes is related to the poor energy efficiency of housing and heating infrastructures. The lack of capital investment stems from the broader 'energy efficiency gap', which can be attributed to the slow acceptance of policies and technologies for the more efficient production and use of energy.

Thus, many of the post-socialist countries' structural difficulties in the energy, housing, and social welfare sectors are produced by the inadequate policy understanding of the interdependence between spatial infrastructures and social exclusion. Even though the initial disregard of social safety nets has gradually been overcome in theory, there is little recognition of the importance of capital stocks and energy efficiency in the design of social policies. Most countries have yet to develop comprehensive programmes to support housing investment among low-income households. The strategic-relational nature of energy poverty implies that a comprehensive solution to the problem will require a simultaneous improvement of the co-ordination and capacity of the relevant institutions operating in the energy, social security, housing and health sectors. There is a need for creating a system of compensatory social welfare measures, to address the salient economic, housing and energy efficiency problems of the energy poor.

The institutional-spatial multiplication of energy poverty

The policies that have led to the emergence of domestic energy deprivation are rooted in the institutional networks of governmental institutions, as well as the distribution and composition of spatial infrastructures. The legacies of antecedent political systems are reflected in both the decision-making process in the energy, social and housing sectors, as well as the nature of the built environment. Institutions and space play a key role in determining whether and to what extent a household may be energy poor.

The reviewed evidence points to three mechanisms for the institutional production of energy poverty in Macedonia and the Czech Republic. First, the 'strategic selectivity' of central governments drives the growing gap between energy prices and stagnant incomes in post-socialism. The strategies of the central state have promoted quick energy price liberalization, at the expense of adequate social protection and demand-side energy efficiency. This discrepancy has led to the emergence of domestic energy deprivation, which is further reinforced by the policies of international financial institutions, and the weak capacity of local government.

Second, both countries possess an 'institutional trap' at the boundaries of their energy, housing and social welfare sectors. The Macedonian institutional trap is significantly larger, because it envelops the entire energy sector. Although the state has formally adopted restructuring measures in line with international demands, governing elites have effectively been interested only in increasing energy prices, and preserving the political and financial benefits of distorted investment decisions. Together with similar processes in the housing and social welfare sectors, this situation reproduces the systemic legacies of socialism, while creating new structural problems typical of the post-socialist period. Conversely, the Czech institutional trap is more limited, although it too functions at the boundaries of the energy, social and rental housing sectors. Its key component is the use of rent control as an across-the-board social protection mechanism – an inadequate instrument for the protection of poor households, due to lack of proper targeting, and the distorting effect on property relations. The problem is aggravated by the disconnection between energy efficiency and housing subsidies, as well as the fact that the state housing allowance amounts to a general poverty relief programme, rather than a specialized safety net.

The third institutional driving force of energy poverty is the 'energy efficiency gap', which stems from the poor integration of energy saving technologies in residential energy distribution and consumption. The efficiency gap is rooted in the legacies of socialism, which did not see energy as a service associated with a set cost. Macedonian decision-makers and entrepreneurs are still grappling with the efficiency dimension of household energy use, as of yet there have been no attempts to move towards less energy-intensive housing policies. The energy efficiency gap in the Czech case is reinforced by conflict of culture and interest between governmental institutions. At the same time, socially-disadvantaged and -marginalized populations continue to suffer due to the stalled reforms in the fiscal system, and the rental housing sector.

The energy efficiency gap is a crucial component of social inequality and deprivation, because it increases the cost of delivered warmth in the home. The fact that some dwellings are less efficient than others adds another layer of poverty on top of the existing patterns of socio-spatial segregation. In addition, the inefficient use of energy in households has serious environmental implications, due to the high amount of carbon emissions and nuclear waste associated with electricity and heat generation in post-socialism. Increased residential energy efficiency is a 'win-win' solution: if states paid more attention to capital investment in low-income housing, they could achieve a dual effect of improving both the economic and environmental quality of life for the socially-excluded.

Having reviewed the role of the social, natural and built environment in the production of energy poverty, I found that these institutional mechanisms are reinforced by analogous socio-spatial processes. The lack of adequate redistributive policies for rising energy prices, stemming in part from the strategic selectivity of the state, has rendered some households more disadvantaged than others. Vulnerability to energy poverty is conditional on a set of economic, demographic and spatial factors, including: the size and source of income, family composition, and the type and quality of housing. This implies that the underlying contingencies of energy deprivation extend beyond the simple ratio of income and prices. The phenomenon of domestic energy deprivation can thus be linked to a wider array of relations in the social and housing sectors, notwithstanding the fact that even the 'simple' ratio of incomes and prices is expressive of deeper economic and political trends. This means that energy poverty is not correlated to relative or absolute poverty lines, although it does affect the households in the lowest income bands. The difference in the demographic extent of the problem in Macedonia and the Czech Republic shows that state policies in non-wage-related sectors can play an important role in promoting and/or reducing domestic energy deprivation.

In spatial terms, the energy efficiency gap in the residential energy sector has been found to comprise two different types of discrepancies, the first of which refers to the difference between the 'raw' energy arriving at the perimeter of the home, on the one hand, and the 'final' warmth reached inside it, on the other. The second component is embodied in the discrepancy between the consumption of housing and energy among some deprived households, whose higher poverty headcounts are due to the use of unaffordable or inflexible heating systems. Taken together, these two sets of relations lead to a paradox whereby the poor pay a higher price per unit of warmth. Thus, even though some of the households suffering from the consequences of the energy efficiency gap may have incomes above the relative or absolute poverty line, they are nevertheless living in energy poverty because they are paying a higher price for each unit of warmth delivered to their home, due to the low efficiency of the housing structure, or the inability to switch to more flexible/affordable fuels. This situation embodies the relational nature of domestic energy deprivation, and the linkages between the institutional and spatial production of the problem.

In general, it can be said that the underpinnings of energy poverty reside in the decreased capabilities of socially-excluded families and individuals in response to the post-socialist transformation process. Many households are deprived of adequate warmth because they are unable to use their housing and intellectual capital as a means of responding to transformation-induced societal change.

The 'messiness' of economic spaces

The unravelling of energy poverty in post-socialism is consistent with the recent emphasis of the 'relational' and 'institutional' in economic geography. The reviewed empirical information about the organizational/spatial production of insufficient domestic warmth supports the argument about 'the world that is more concrete, institutionally cluttered, and – perhaps above all – geographically uneven' (Peck, 2000, p. 61). Domestic energy deprivation interacts with culture and economy in a manner which demonstrates 'the importance and diversity of culturally embedded non-market, non-capitalist processes that underpin household economic practices' (Smith, 2002: 234).

Aside from providing further empirical backing for existing theoretical arguments, however, the reviewed evidence has generated several new sets of findings, which can be developed into broader directions and systems. For example, I have concluded that post-socialist poverty cannot – and must not – be reduced to incomes and prices, as is often advocated among some systems of thought. Qualitative and quantitative data alike show that social exclusion is a product of multiple layers of institutional interests and interactions, which as such are path-dependent and embedded in antecedent economic substrata. What is more, each institutional mechanism is matched by an analogous process in space, whose emergence can be attributed partly to the institutional process itself, and partly to the legacies of past housing, demographic and development policies.

Moreover, it has emerged that energy poverty exists at the nexus of institutional and spatial processes, within a gap that lies beyond the mainstream focus of energy reforms, social safety nets, and housing investment. Taken together, however, the institutional and spatial aspects of energy poverty are a path-shaping phenomenon, because they have the power to perpetuate a desired set of organizational and architectural arrangements. There is a wide body of empirical evidence to support this mutual reinforcement within the specific cases of Macedonia and the Czech Republic, and in the broader regional context.

At a more fundamental level, an argument can be made about the 'messiness' of economic spaces, the wider implication of which is that current macro- and micro-economic models of energy and housing consumption cannot capture the full set of processes surrounding structural change of the economy. The problem with policies based on 'neat' macro- and micro-economic models is that overlooked dynamics accumulate and strengthen each other over time, to the point when it is impossible to relegate them under *ceteris paribus*. In many socialist countries, this situation has stalled structural reforms, rendering the argument about the inter-separability of equity and efficiency meaningless. It is thus necessary to move towards a more integrated understanding of social exclusion, whereby this situation is seen as the result of reduced capabilities and utility at the household level.

The need for incorporating evolutionary and institutional understandings of economic change into mainstream policy reforms is exemplified by the emergence of the institutional trap in the Czech rental sector, which is grounded in the spatial economies of energy and housing consumption of disadvantaged households. The disproportionately high housing costs of disadvantaged strata have been neglected within most mainstream policy proposals for rent deregulation. This could be due to the weak policy understanding of the power-geometries of the institutional trap, as it is still inadequately recognized that the roots of social exclusion and deprivation are interwoven within the boundaries of housing, energy and social welfare policies. Addressing the difficult housing circumstances faced by the majority of pensioners, as well as single parent and multi-children families, requires concerted efforts to address the chronic lack of capital investment and spatial flexibility in the housing stock that they occupy. Otherwise, an even greater part of the population will fall into poverty, as a result of the removal of rent control.

The book has also endeavoured to demonstrate that future research of social exclusion and systemic reforms should incorporate a balanced mix of qualitative and quantitative methods. While it is impossible to situate energy poverty within the broader demographic and economic context without executing statistical analyses of nationally-representative data sets, these tools are, in themselves, insufficient for examining the everyday production of inequality in the grain of society. In-depth interviews and biographies can fill many of the analytical gaps left behind by statistics.

Landscapes of social exclusion in the 'new Europe'

In addition to contributing to broader theoretical debates within economic geography, the reviewed data provides important insights into the emergent geographies of inequality in the European continent as a whole. The variation of energy poverty factors across different countries or groups of countries, shows that the 'new' patterns of social inequality and poverty in Central and Eastern Europe are organized along territorial lines that reflect deeper historical features, such as institutional cultures, choice of economic system, and household-level responses to poverty.

This discrepancy arises out of the plurality of reform paths in different post-socialist countries, where the economic and political transformation process has brought increased economic divergence (see, for example, Dunford and Smith, 2000). However, Chapter 6 established that the geographies of difference do not end at the national level, as the spaces of energy deprivation also vary at smaller scales, ranging from meso-level regions, up to individual apartments. It was found that regional-level economic, physical and housing disparities may play a key role in determining the 'geographies' of energy poverty. While supporting the argument about new form of uneven development initiated by the post-socialist transformation (Bradshaw and Stenning, 2000; Smith and Pickles, 1998), this conclusion also raises additional questions about the role of scale in the production of social exclusion.

At a more practical level, the findings of the book imply that some types of urban and rural settlements in Central and Eastern Europe may be more susceptible to social exclusion than others. In the Czech Republic (and possibly, other Central European countries) some peripheral socialist panel estates, and pre-war tenement buildings in inner cities, are suffering from a lack of maintenance and capital investment. Macedonia is facing a similar predicament, with the added problem of inadequate maintenance of socialist housing estates, and substandard suburban and rural housing. The reviewed evidence suggests that such problems also exist in other Balkan and post-Soviet states. In all of these cases, disproportionate energy burdens may exacerbate existing poverty traps, by disabling households from investing in the upkeep and refurbishment of their dwellings.

The poverty of flexibility

Many of the disadvantaged households interviewed for the purposes of this study have been facing exorbitantly high energy costs as a result of living in houses and/or using heating systems that did not match their needs. In the cases of these families, the emergence of domestic energy deprivation is connected to the inability to switch away from an unsustainable housing arrangement. It can be said that they are suffering from a poverty of 'flexibility', where deprivation emerges as a result of the inability of the household to adapt to new circumstances – in this case, increased energy prices and reduced welfare support related to the overall dismantling of the socialist welfare state, and the neoliberalization of the energy sector.

The *Oxford American Dictionary* defines the property of 'flexibility' as the 'ability to bend easily without breaking'. Thus, a flexible object is 'one that is able to respond to altered circumstances or conditions', while a flexible person is 'ready and able to change so as to adapt to different circumstances'. Following Mandelbaum's (1978) definition of flexibility as 'the ability to respond effectively to changing circumstances', Jonsson (2006) develops a more elaborate understanding of the concept, whereby it is seen as 'the propensity of an actor or a system to exhibit variation in activities or states which is correlated with some other variation and desirable in view of this variation' (p. 2). He points out that 'adaptive change of activities occurs as a result of (a) ability to change, (b) willingness to change and (c) changing circumstances in view of which change of activities becomes desirable' (ibid). However, even though this definition implies that flexibility is desirable *per se*, he is quick to point out that 'variability regarded as flexibility may be 'bad' even for an actor for which this variability is desirable if its consequences are disregarded' (ibid). It is partly because of such ambiguities that the concept of flexibility has taken very different meanings and interpretations across different socio-economic contexts, while its usefulness has been contested by a wide range of authors (see, for example, Pollert, 1988; 1991; Sennett, 1998).

For some of the interviewed households, the inflexibility of the residential environment means that they have been pushed into energy poverty. The 'relational' (or 'topological') remoteness of such families from the remainder of society is increasing, although the 'Euclidean' ('topographic') distance remains the same. They are having to face the continued persistence of severe, often banal, temporal and spatial constraints on everyday life and socio-economic change (Kirsch, 1995), although more affluent residents enjoy easy access to facilities and amenities as a result of their ability to pay for improved energy and housing services. Thus, 'flexibility' has very different meanings for different population groups in post-socialism.

Fighting energy poverty

The analysis of the structural causes of energy poverty (in Chapters 3, 4, 5 and 6) can be used as a basis for proposing practical policy measures for its resolution. However, the context-specific nature of domestic energy deprivation implies that each country will have to implement a unique set of energy poverty alleviation strategies. Probably the only recommendation valid for the entire post-socialist region is the need to develop a clear and direct energy poverty measurement framework across Europe. This implies that household expenditure surveys, which are the main instrument for defining poverty lines, would have to involve the gathering and publication of a wider range of household data. In particular, they must contain income, expenditure, energy consumption data and housing quality information that is inter-comparable for different groups. In practical terms, this would require increasing the capacity of local statistical and scientific organizations, perhaps with material support from international financial institutions.

Collecting and disaggregating housing and expenditure data according to household income and type would allow vulnerable households to be identified on the basis of the relationship between income, housing and the cost of warmth. This could lead to a common, trans-European methodology for energy poverty measurement, extending beyond mere affordability calculations. An approach like this could help estimate the size and extent of populations affected by energy poverty by interpolating a number of different indicators, including actual energy consumption, expenditure, demographic structure and the state of the dwelling. Having a clear picture of the size of the energy poverty problem (perhaps through a composite energy poverty 'index') may also lead to an increased political awareness about the issue, as it will underline its far-reaching effects on the economy. Such an approach could also be implemented outside Europe, especially in the global South where energy poverty is an even bigger problem.

Beyond this universal step, however, energy poverty alleviation policies would have to take different paths in the different post-socialist states. In countries like Macedonia, where energy poverty is pervasive, there is a need to remove households from the grasp of the institutional trap and the efficiency gap in the energy and social welfare sectors. It will be important to insist that the necessary degree of consumer protection and demand-side efficiency management is incorporated in energy legislation, regardless of the degree of private capital in the energy sector. In addition, international organizations and NGOs alike will have to continue to lobby for the implementation of a wider range of energy efficiency measures, including mandatory standards and labelling. The social welfare sector is also in need of reforms: a well-targeted social safety net, incorporating lifeline tariffs and/or capital investment, could provide an adequate response to future energy price increases. It must also be ensured that cheap fuelwood is available and accessible to all, as this type of energy has played a progressive role in reducing energy poverty. Last but not least, Macedonia should consider the adoption of a formal housing policy document, and the establishment of a separate government agency for support of low-income housing and disadvantaged regions, much like the Czech Ministry for Regional Development. Many of the country's current economic and political problems (including the marginalization of ethnic minority populations) stem from the absence of a concerted state strategy to curb uneven development and socio-spatial exclusion.

Although the Czech Republic's problems are of a comparatively lesser magnitude, this country should also have to implement a number of brave policy steps, if it is to move towards more socially just and environmentally sustainable patterns of energy consumption. First, the state must untangle the rent deregulation gridlock, by speeding up the liberalization of rents and owner-tenant relations, and providing the necessary income support for the households who need it. Second, the housing allowance must be reformed to account for household-level energy and housing burdens, possibly in line with the current German model (a detailed proposal has already been developed by the Czech Sociological Institute, see Sunega, 2001). Third, there is a need to increase the capacity and funding of nascent government departments such as the Ministry of Environment and the Ministry for Regional Development, so that they may fulfil their intention of supporting the transformation of the efficiency markets, and increasing/sharpening housing subsidies for low-income households. The fact that many such reforms are already underway attests to the greater institutional strength of policy frameworks in this country.

At a more general level, it is important to stress that relying solely on instrumentalized methods for poverty alleviation (such as disbursing subsidies on the basis of a dualistic division of the population into 'poor' and 'non poor', as advocated in Lovei et al., 1999) may not deliver the desired results. The continuing reliance on such mechanisms has the potential to bring severe hardship for the majority of the population in the 'lagging' transformation states, where widespread income poverty will be exacerbated by rapid tariff increases. It will be necessary to convince governing elites to embrace the long-term benefits of integrated energy poverty alleviation measures, rather than maintaining the *status quo* or promoting the strategic selectivity of the state. As pointed out, among others, by Velody et al. (2003), energy poverty support is most effective when it involves an energy efficiency component to address the salient loss of effective warmth in the homes of vulnerable households.

Neoliberalism, globalization, and the post-welfare state

Last but not least, the reviewed evidence raises several wider questions about the nature of globalization in the former Communist states of Eastern Europe. It has become clear that post-socialist structural reforms cannot be abstracted from broader economic and political trends, because globalization shapes the content and direction of the transformation process. This is demonstrated by the widening institutional gap between Macedonia and the Czech Republic, which are representative of supra-national groups of countries. Such empirical insights feed into broader theoretical debates about the importance of space and place in the post-socialist transformation. Most notably, it remains unclear whether the generalized regional patterns of energy poverty be used as a basis for arguing that globalization is 'overwhelming local communities and nation-states' (something negated, for instance, in Boyer and Drache, 1996). Or perhaps the structural differences between Macedonia and the Czech Republic may be indicative of the contrary, because the legacies of past economic and political systems, combined with micro-level knowledge and coping strategies, are the key factor of restructuring choices taken in a particular situation? There is evidence to support both lines of thought, which means that further research and discussion is needed, taking into account the inter-relationships of these processes within different temporal and spatial contexts.

Moreover, it appears that most countries simply have no policy choice other than to follow a pre-determined reform path in the energy, housing and social welfare sector. In general, this comprises energy liberalization, social welfare reduction, and the delegation of housing and social policies to non-state actors. Such processes is aided by a complex synergy of global power and central governments, which have strived to turn neoliberalism into an inevitable path for the post-socialist transformation. Its principles are being gradually integrated into the legal systems of

ECE and FSU states, making it increasingly difficult to formulate alternative modes of economic regulation.

Thus, although the post-socialist countries never had a Fordist regulation to begin with, they are becoming part of the global post-Fordist regime of accumulation. Both Macedonia and the Czech Republic (and, according to the 'transition' literature, many other post-socialist countries) are characterized by a process typical to the post-Fordist welfare state, namely the 'transition towards a new regime of accumulation built upon 'flexibility'', in which 'the state's provision of welfare will be quite different' (Eatwell et al., 2000, p. 21, also see Pierson, 1999). This 'post-welfare' state is less generous to all but the most-needy groups (defined by a set of formal criteria), and is disbursing its services via new types of institutional arrangements. In the more 'reformist' countries, there is a tendency to transfer social service provision to third parties, such as civil society, while the remainder of the region is trying to find a capitalist fit for the old system of indirect social support (implicit subsidies, enterprise support, and so on). Both trends are point to the decomposition of the old regime of social welfare under a neoliberalizing project, similar to the situation in the rest of the non-socialist world.

The reduction of income support to a limited set of social groups raises important ethnical and human rights issues, among other problems. Article 25 of the Universal Declaration of Human Rights states that 'everyone has the right to a standard of living adequate for the health and well-being of himself *(sic)* and of his family, including food, clothing, housing and medical care and necessary social services' (UNGA, 1985). Although affordable energy is not mentioned explicitly in this text, the emphasis on 'health' and 'housing' could imply that the 'right to be warm' is, too, a basic human right, and the policies that reduce the affordability of energy are in default of the Declaration.

Endnote

Aside from describing and analysing a hitherto-unknown form of social inequality, this book has also had a deeper purpose: to advance a fundamental theoretical argument about the importance of space in social science. I have argued in favour of integrating institutions and space in analyses of social exclusion. My principal claim is that we can understand and interpret poverty only by embracing the multiplicity of its underlying organizational and spatial processes. This can lead not only towards more effective poverty policies, but also an improved understanding of the dynamics that create exclusion and deprivation more generally.

Bibliography

Adam, J. (1996), 'Social costs of transformation in the Czech Republic', *Moct-Most: Economic Policy in Transitional Economies* 6, 163–183.

Adam, J. (1999), *Social Costs of Transformation to a Market Economy in Post-Socialist Countries: The Case of Poland, the Czech Republic and Hungary* (Basingstoke, Macmillan).

ADHCR (Association for District Heating of the Czech Republic) (published online in 2007), 'Energetika v tabulkách a grafech [The energy sector in tables and graphs]', <http://www.tscr.cz/cz/grafy/grafy.htm>, accessed 20 February, 2005.

AEA (Austrian Energy Agency) (published online in 2006), 'Energy in Central and Eastern Europe', <http://www.eva.ac.at/enercee>, accessed 9 October 2006.

Albeda, R. and Withorn, A. (eds) (2002), *Lost Ground: Welfare Reform, Poverty and Beyond* (Boston, South End Press).

Alexandrova, G. (2000), 'NEC is no longer monopolist on the energy market', *Capital* 4, 12–15.

Allsopp, C. and Kierzkowski, H. (1997), 'The assessment: economics of transition in Eastern and Central Europe', *Oxford Review of Economic Policy* 13, 1–22.

Amin, A. (1999), 'Placing globalisation', in Bryson et al. (eds).

Amsden, A. et al. (1994), *The Market Meets its Match: Restructuring the Economies of Eastern Europe* (Cambridge, Mass., Harvard University Press).

Anderson, P.W. et al. (eds) (1988), *The Economy as an Evolving Complex System* (Santa Fe, Addison-Wesley Publishing Company).

Anderson, R. (2001), 'Privatisation clock ticking', *Financial Times Energy and Utilities Review/Central Europe* August 2001, 2.

Andrusz, G. et al. (eds) (1996), *Cities After Socialism: Urban and Regional Change and Conflict in Post-Socialist Societies* (Oxford, Blackwell).

Anex, R.P. (2002), 'Restructuring and privatizing electricity industries in the Commonwealth of Independent States', *Energy Policy* 30, 397–408.

Arpaillange, J. (1995), *The Buildings and the People: An Integrated Approach to Apprehend Energy-Housing Issues in a Former Soviet Republic* (Washington, D.C., World Bank).

Arthur, W.B. (1988), 'Self-reinforcing mechanisms in economics', in Anderson et al. (eds).

Åslund, A. (1992), *Post-Communist Economic Revolutions: How Big a Bang?* (Washington, D.C., Centre for Strategic and International Studies).

Åslund, A. (1994), 'The case for radical reform', *Journal of Democracy* 5, 63–74.

Åslund, A. (2002), *Building Capitalism: The Transformation of the Former Soviet Bloc* (Cambridge, Cambridge University Press).

Åslund, A. et al. (2002), 'Escaping the under-reform trap', *IMF Staff Papers* 48, 88–108.

Atkinson, A.B. and Hills, J. (1998), *Exclusion, Employment and Opportunity* (London, Centre for Analysis of Social Exclusion).

Atkinson, A.B. and Micklewright, J. (1991), *Economic Transformation in Eastern Europe and the Distribution of Income* (Cambridge, Cambridge University Press).

Atkinson, A.B. et al. (1995), *Income Distribution in OECD Countries*, OECD Social Policy Studies No. 18. (Paris, OECD).

Avdiushin, S. (1997), *Climate Change Mitigation: Case Studies From Russia* (Richland, PNNL (Pacific Northwest National Laboratory)).

Bacon, R. (1995), *Measurement of Welfare Changes Caused by Large Price Shifts*, World Bank Discussion Paper No. 273. (Washington, D.C., World Bank).

Balmaceda, M.M. (ed.), (2002), *Forschungsschwerpunkt Konflikt-und Kooperationsstrukturen in Osteuropa [Research Project on Conflict and Co-operation Structures in Europe]* (Manheim, University of Manheim).

Balmaceda, M.M. (2002), 'EU energy policy and future european energy markets: consequences for the Central and East European States', in Balmaceda (ed.),

Barlow, J. (1993), 'Reforming the housing system in Eastern Europe', *Policy Studies* 14, 53–62.

Baross, P. and Struyk, R. (1993), 'Housing transition in Eastern Europe', *Cities* 10, 179–188.

Barr, N. (1996), 'Income transfers in transition: constraints and progress', *Moct-Most: Economic Policy in Transitional Economies* 6, 57–74.

Barr, N. (1998), *The Economics of the Welfare State* (Oxford, Oxford University Press).

Bentham, J. (1996), *An Introduction to the Principles of Morals and Legislation* (Oxford, Clarendon Press).

Bhalla, A. and Lapeyre, F. (1997), 'Social exclusion: towards an analytical and operational framework', *Development and Change* 28, 413–433.

Bijker, W. and Law, J. (eds) (1992), *Shaping Technology, Building Society: Studies in Sociotechnical Change* (Cambridge, M.A., MIT Press).

Bingham, N. (1996), 'Object-ions: from technological determinism towards geographies of relations', *Environment and Planning D: Society and Space* 14, 635–657.

Blake, E. (2001), 'Spatiality past and present: an interview with Edward Soja, Los Angeles, April 2001', *Journal of Social Anthropology* 2, 139–158.

Blanchard, O. et al. (1992), *Reform in Eastern Europe* (London, MIT Press).

Boardman, B. (1991), *Fuel Poverty: From Cold Homes to Affordable Warmth* (London, Bellhaven).

Bodnár, J. (1996), 'He that hath to him shall be given: housing privatization in Budapest after state socialism', *International Journal of Urban and Regional Research* 20, 616–636.

Bonnefoy, X. and Sadeckas, D. (2006), 'A study on the prevalence, perception, and public policy of 'fuel poverty' in European countries', Paper presented at the workshop on 'Joint action to combat energy poverty in Europe: research and policy challenges', 27 September 2006, Oxford.

Boschma, R.A. and Lambooy, J.A. (1999), 'Evolutionary economics and economic geography', *Journal of Evolutionary Economics* 9, 411–429.

Boušová, K. (published online in 2006), 'Deregulace nájemného schválena, trh s bydlením se konečně začne uvolňovat! [Deregulation of rent adopted, the housing market will finally start opening up!]', <http://www.penize.cz/info/zpravy/zprava. asp?IDP=1andNewsID=4237>, accessed 9 October 2006.

Boyer, C. (1995), 'The great frame up: fantastic appearances in contemporary spatial politics', in Liggett and Perry (eds).

Boyer, R. and Drache, D. (eds) (1996), *States Against Markets: The Limits of Globalization* (London and New York, Routledge).

Braber, M. (2000), 'Estimations of poverty lines in Macedonia', in Hutton (ed.),

Braber, R. and van Tongeren, F. (1996), 'Energy price reforms in Russia', *Most-Most* 6, 139–162.

Bradshaw, J. and Harris, T. (eds) (1983), *Energy and Social Policy* (London, Routledge and Kegan Paul).

Bradshaw, M.J. (ed.), (1997), *Geography and Transition in the Post-Soviet Republics* (Chichester, Wiley).

Bradshaw, M.J. and Kirkow, P. (1998), 'The energy crisis in the Russian Far East: origins and possible solutions', *Europe-Asia Studies* 50, 1043–1064.

Bradshaw, M.J. and Stenning, A. (2000), 'The progress of transition in East Central Europe', in Bachtler et al. (eds).

Brathwaite, J. et al. (2000), *Poverty and Social Assistance in Transition Countries* (Basingstoke, Macmillan).

Bryson, J. et al. (eds) (1999), *The Economic Geography Reader: Producing and Consuming Global Capitalism* (New York, Wiley).

Buchanan, J.M. (1975), *The Limits of Liberty: Between Anarchy and Leviathan* (Chicago, University of Chicago Press).

Buckley, R.M. et al. (2001), *Urban Housing and Land Market Reforms in Transition Countries: Neither Marx Nor Market* (Washington, D.C, World Bank).

Buckley, R.M. and Gurenko, E.N. (1997), 'Housing and income distribution in Russia: Zhivago's legacy', *The World Bank Observer* 12, 19–32.

Burawoy, M. and Verdery, K. (eds) (1999), *Uncertain Transition: Ethnographies of Change in the Postsocialist World* (Lanham, Rowman and Littlefield).

Burdett, M. (ed.), (2002), *Central and Eastern Europe and FSU Electricity Prospects for 2002* (London, Platts).

Burghes, L. (1980), *Living From Hand to Mouth: A Study of 65 Families Living on Supplementary Benefit* (London, CPAG).

Buzar, S. (2005), 'The institutional trap in the Czech rental sector: nested circuits of power, space and inequality', *Economic Geography* 82, 381–405.

Buzar, S. (2006a), 'Estimating the extent of domestic energy deprivation through household expenditure surveys', *CEA Journal of Economics* 1, 5–14.

Buzar, S. (2006b), 'Energy poverty in Macedonia and the Czech Republic', *Beyond Transition* 17, 15–17.

Buzar, S. (2007a), 'The "hidden" geographies of energy poverty in post-socialism: between institutions and households', *Geoforum* 38, 224–240.

Buzar, S. (2007b), 'Energy and environment problems in the Western Balkans: an overview of the EBRD's response', *Geografiska Annaler B*, forthcoming.

Buzar, S. (2007c), 'When homes become prisons: the relational spaces of post-socialist energy poverty', *Environment and Planning A*, 1908–1925.

Byrne, D. (1999), *Social Exclusion* (Buckingham, Open University Press).

Carter, F.W. and Turnock, D. (eds) (1993), *Environmental Problems in Eastern Europe* (London, Routledge).

Caselli, G.P. and Battini, M. (1997), 'Following the tracks of Atkinson and Micklewright: the changing distribution of income and earnings in Poland from 1989 to 1995', *Moct-Most: Economic Policy in Transitional Economies* 7, 1–19.

Castells, M. (2002), *The Rise of the Network Society* (Oxford, Blackwell).

CDR (Centre for Development Research) (1999), *Local Organisations and Rural Poverty Alleviation. Project Reports* (Copnehagen, CDR).

CEEBIC (Central and Eastern Europe Business information Center) (1997), *Czech Energy Market* (Washington, D.C., CEEBIC).

Clapham, D. and Kintrea, K. (1996), 'The patterns of housing privatization in Eastern Europe', in Clapham (ed.),

Clark, G.L. et al. (2000), 'Economic geography: transition and growth', in Clark et al. (eds).

Clark, G.L. et al. (eds) (2000), *The Oxford Handbook of Economic Geography* (Oxford, Oxford University Press).

Commander, S. and Mumssen, C. (1998), *Understanding Barter in Russia*, EBRD Working Paper No. 37. (London, EBRD).

Cornia, G.A. (1994), 'Income distribution, poverty and welfare in transitional economies: a comparison between Eastern Europe and China', *Journal of International Development* 6, 569–607.

Cornia, G.A. et al. (1996), 'Policy, poverty and capabilities in the economies of transition', *Moct-Most: Economic Policy in Transitional Economies* 6, 149–172.

Crnobrnja, M. (1991), 'Yugoslavia's energy choices and the economic dimension', in DeBardeleben (ed.),

CSD (Centre for Social Development) (2005), *Impact of Utility Charges on Poor Households Survey Report* (Ulanbataar, CSD).

CSO (Czech Statistical Office) (1996), *Práce, Sociální Statistiky: Životní Úroveň [Employment and Social Statistics: Living Standards]* (Prague, CSO).

CSO (Czech Statistical Office) (2002), *Práce, Sociální Statistiky: Životní Úroveň [Employment and Social Statistics: Living Standards]* (Prague, CSO).

CSO (Czech Statistical Office) (2005), *Příjmy Domácnosti a Míra Chudoby v ČR [Household Incomes and Poverty Rates in the Czech Republic]* (Prague, CSO).

CSO (Czech Statistical Office) (2006), *Práce, Sociální Statistiky: Životní Úroveň [Employment and Social Statistics: Living Standards]* (Prague, CSO).

Dallago, B. (1999), 'Convergence and divergence of economic systems: a systemic-institutional perspective', *Economic Systems* 23, 167–172.

Davidson, P. and Kregel, J.A. (eds) (1997), *Improving the Global Economy: Keynesianism and the Growth in Output and Employment* (Cheltenham, Edward Elgar).

de Certeau, M. (1984), *The Practice of Everyday Life* (Berkeley, University of California Press).

de Deken, J.J. (1994), *Social Policy in Postwar Czechoslovakia: The Development of Old-Age Pensions and Housing Policies During the Period 1945–1989* (Florence, European University Institute).

DeBardeleben, J. (ed.), (1991), *To Breathe Free: Eastern Europe's Environmental Crisis* (Washington, D.C, Baltimore, London, Woodrow Wilson Center Press, and Johns Hopkins University Press).

Dnevnik (2002), 'Vo penzionerskiot dom Čair, stanarite so mali penzii se ili gladni ili im e studeno [The low-income inhabitants of the Čair nursing home are either hungry or cold]', 20 November 2002.

Dnevnik (2007), 'Ekspertite predupreduvaat na energetska kriza [Experts warn of energy crisis]', 20 February, 2007.

Dodonov, B. et al. (2001), *How Much Do Electricity Tariff Increases in Ukraine Hurt the Poor?* (Kiev, Institute for Economic Research and Policy Consulting.).

Dodonov, B. et al. (2004), 'How much do electricity tariff Increases in Ukraine hurt the poor?' *Energy Policy* 32, 855–863.

Duke, V. and Grime, K. (1997), 'Inequality in post-Communism', *Regional Studies* 31, 883–890.

Dunford, M. (1998), 'Differential development, institutions, modes of regulation and comparative transitions to capitalism: Russia, the Commonwealth of Independent States and the Former German Democratic Republic', in Pickles and Smith (eds).

Dunford, M. and Smith, A. (2000), 'Catching up or falling behind? Economic performance and regional trajectories in the "new" Europe', *Economic Geography* 76, 169–195.

EABE (European Academy of the Built Environment) (1996), *Environmental Improvements in Prefabricated Housing Estates* (EABE, Berlin).

Eatwell, J. et al. (2000), *Hard Budgets, Soft States: Economics of Social Policy Choices in Central and Eastern Europe* (London, Institute for Public Policy Research).

EBRD (European Bank for Reconstruction and Development) (1999), *Energy Operations Policy* (London, EBRD).

EBRD (European Bank for Reconstruction and Development) (2000a), *Transition Report 2000: Employment, Skills and Transition* (London, EBRD).

EBRD (European Bank for Reconstruction and Development) (2000b), *Hungary Investment Profile 2001* (London, EBRD).

EBRD (European Bank for Reconstruction and Development) (2000c), *Latvia Investment Profile 2001* (London, EBRD).

EBRD (European Bank for Reconstruction and Development) (2001), *Transition report 2001: Energy in Transition* (London, EBRD).

EBRD (European Bank for Reconstruction and Development) (2003), *Can the Poor Pay for Power? The Affordability of Electricity in South East Europe* (London, EBRD and IPA Energy).

EBRD (European Bank for Reconstruction and Development) (2005), *Transition report 2005: Business in Transition* (London, EBRD).

Economics Glossary (published online in 2006), 'Compensating Variation', <http://economics.about.com/library/glossary/bldef-compensating-variation.htm>, accessed 1st October 2006.

Ellman, M. (1994), 'Transformation, depression, and economics: some lessons', *Journal of Comparative Economics* 19, 1–21.

Enyedi, G. (1990), 'Specific urbanization in East-Central Europe', *Geoforum* 21, 163–172.

EPCM (Electric Power Company of Macedonia) (2003), *Annual Report* (Skopje, EPCM).

ESI (European Stability Initiative) (2002), *Ahmeti's Village. The Political Economy of Interethnic Relations in Macedonia* (Skopje and Berlin, ESI).

Euroheat (published online in 2006), 'Country-by-country statistics', <http://www.euroheat.org/redirects/redirect_cbyc.htm>, accessed 10 October 2006.

Evans, M. (1995), *Energy Efficiency Business Opportunities in Ukraine*, Report on the Mission of the U.S.-Ukrainian Expert Working Group on Energy-Efficiency Options for Closing the Chernobyl Reactors. (Richland, Pacific Northwest National Laboratory).

Evans, M. (2000), *Tapping the Potential for Energy Efficiency: The Role of ESCOs in the Czech Republic, Ukraine and Russia*, Report on the Mission of the U.S.–Ukrainian Expert Working Group on Energy-Efficiency Options for Closing the Chernobyl Reactors. (Richland, PNNL (Pacific Northwest National Laboratory)).

Fajth, G. (1999), 'Social security in a rapidly changing environment: the case of the post-Communist transformation', *Social Policy and Administration* 33, 416–436.

Fankhauser, S. and Tepic, S. (2005), *Can Poor Consumers Pay for Energy and Water?*, Working Paper No. 92. (London, EBRD).

FCSRD (Forum Centre for Strategic Research and Documentation) (2002), *Urban Poverty in Macedonia – Background Paper Towards the National Poverty Strategy* (Skopje, FCSRD).

Ferge, Z. (1992), 'Social policy regimes and social structure. Hypotheses about the prospects of social policy in Central-Eastern Europe', in Ferge and Kolberg (eds).

Ferge, Z. and Kolberg, J.E. (eds) (1992), *Social Policy in a Changing Europe* (Boulder, Campus and Westview).

Fischer, S. and Gelb, A. (1991), 'The process of socialist economic transformation', *Journal of Economic Perspectives* 5, 91–105.

Folwell, K. (1999), *Getting the Measure of Social Exclusion* (London, London Research Centre).

Freund, C.L. and Wallich, C.I. (1996), 'The welfare effects of raising household energy prices in Poland', *The Energy Journal* 17, 53–78.

GCR (Government of the Czech Republic) (2001), *Energy Policy* (Prague, GCR).

Gillan, J. (2000), 'Household budget and house conditions in two Russian cities', in Hutton (ed.),

Goss, J. (1988), 'The built environment and social theory: towards an architectural geography', *The Professional Geographer* 40, 392–403.

Götting, U. (1994), 'Destruction, adjustment and innovation: social policy transformation in Eastern and Central Europe', *Journal of European Social Policy* 4, 181–200.

Grabher, G. and Stark, D. (eds) (1997), *Restructuring Networks in Post-Socialism: Legacies, Linkages and Localities* (Oxford, Oxford University Press).

Grabher, G. and Stark, D. (1998), 'Organising diversity: evolutionary theory, network analysis and post-socialism', in Pickles and Smith (eds).

Graham, S. (2000), 'Constructing premium network spaces', *International Journal of Urban and Regional Research* 24, 183–200.

Graham, S. (2000), 'Introduction: cities and infrastructure networks', *International Journal of Urban and Regional Research* 24, 114–119.

Graham, S. and Healey, P. (1999), 'Relational concepts of space and place: issues for planning theory and practice', *European Planning Studies* 7, 623–646.

Gray, D. (1995), *Reforming the Energy Sector in Transition Economies: Selected Experience and Lessons*, World Bank Discussion Paper. (Washington, D.C., World Bank).

GRM (Government of the Republic of Macedonia) (2000), *Poverty Reduction Strategy Paper, Interim Version* (Skopje, GRM).

GTZ (German Technical Co-operation) (2001), *Macedonian-German Project: Energy Efficiency in Buildings* (Skopje, GTZ).

Guyett, S. (1991), 'Environment and lending: lessons of the World Bank, hope for the European Bank for Reconstruction and Development', *New York University Journal of International Law and Politics* 24, 889–919.

Hagenaars, A. et al. (1994), *Poverty Statistics in the Late 1980s: Research Based on Micro-Data* (Luxembourg, Office for Official Publications of the European Communities).

Hampl, M. et al. (1999), *Geography of Societal Transformation in the Czech Republic* (Prague, Charles University).

Hausner, J. et al. (1995), 'Institutional change in post-socialism', in Hausner et al. (eds).

Hausner, J. et al. (eds) (1995), *Strategic Choice and Path-Dependency in Post-Socialism* (Aldershot, Edward Elgar).

Havelková, E. and Valentová, B. (1998), 'Komparatívna analýza bytovej politiky v Slovenskej a Českej Republike v rokoch 1990–1996 [Comparative analysis of housing policy in the Slovak and Czech Republics in the years 1990–1996]', in Potůček and Rodičová (eds).

HCHRRM (Helsinki Committee for Human Rights in the Republic of Macedonia) (2002), *Report on the Human Rights Situation in Macedonia* (Skopje, HCHRRM).

Healy, J. (2003), 'Excess winter mortality in Europe: a cross country analysis identifying key risk factors', *Journal of Epidemiology and Community Health* 57, 784–789.

Healy, J. (2004), *Housing, Fuel Poverty and Health: A Pan-European Analysis* (Aldershot, Ashgate).

Hegedüs, J. and Tosics, I. (1994), 'The poor, the rich and the transformation of urban space: editors' introduction to the special issue', *Urban Studies* 31, 989–993.

Hellman, J.S. (1998), 'Winners take all – the politics of partial reform in postcommunist transitions', *World Politics* 50, 203–234.

HEO (Hungarian Energy Office) (2000), *The New Hungarian End-User Electricity Tariff System* (Budapest, HEO).

Hörschelmann, K. and van Hoven, B. (2003), 'Experiencing displacement: the transformation of women's space in (former) East Germany', *Antipode* 33, 742–760.

Hubbard, P. and Holloway, L. (2000), *People and Place: The Extraordinary Geography of Everyday Life* (Harlow, Prentice Hall).

Huff, C. (2000), *The Logic of Political Constraints and Reform With Applications to Strategies for Privatization*, World Bank Working Paper. (Washington, D.C., World Bank).

Hughes, G. (1991), 'The energy sector and problems of energy policy in Eastern Europe', *Oxford Review of Economic Policy* 7, 77–98.

Hulchanski, D. (1995), 'The concept of housing affordability: Six contemporary uses of the housing expenditure-to-income ratio', *Housing Studies* 10, 471–492.

Hutton, S. (ed.), (2000), *Poverty in Transition* (London, Routledge).

Hutton, S. et al. (2000), 'Albania and Macedonia: transitions and poverty', in Hutton (ed.),

IEA (International Energy Agency) (2000), 'Hungary', in IEA (ed.),

IEA (International Energy Agency) (2002), *Energy Policy in the Czech Republic* (Paris, IEA).

IEA (International Energy Agency) (2003), *Energy Policies of IEA Countries* (Paris, IEA).

ILO (International Labour Organization) (1995), *World Labour Report 1995* (Geneva, ILO).

Isherwood, B.C. and Hancock, R.M. (1979), *Household Expenditure on Fuel: Distributional Aspects* (London, Economic Advisor's Office).

ISPJR (Institute for Social Political and Juridical Research) (2002), *Housing Quality in the Republic of Macedonia* (Skopje, ISPJR).

ISPO (Institute for Spatial Planning – Ohrid) (1982), *Prostoren Plan na Socijalistička Republika Makedonija [Spatial Plan of the Socialist Republic of Macedonia]* (Ohrid, ISPO).

Jaffe, A.B. and Stavins, R.N. (1994), 'The energy efficiency gap. What does it mean?' *Energy Policy* 22, 804–810.

Jančar-Webster, B. (1993), 'Yugoslavia', in Carter and Turnock (eds).

Jarvis, H. et al. (2001), *The Secret Life of Cities: The Social Reproduction of Everyday Life* (Harlow, Prentice Hall).

Jasinski, P. and Pfaffenberger, W. (2000), *Energy and Environment: Multiregulation in Europe* (Aldershot, Ashgate).

Jessop, B. (2001), 'Institutional re(turns) and the strategic – relational approach', *Environment and Planning A* 33, 1213–1235.

Johnson, S. et al. (1997), 'The unofficial economy in transition', *Brookings Papers on Economic Activity* 2, 159–239.

Jones, C. and Revenga, A. (2000), *Making the Transition Work for Everyone: Poverty and Inequality in Europe and Central Asia* (Washington, D.C., World Bank).

Jones, M.R. (1997), 'Spatial selectivity of the state? The regulationist enigma and local struggles over economic governance', *Environment and Planning A* 29, 831–864.

Jonsson, D. (2006), 'Flexibility and stability in working life', in Furåker et al. (eds).

Jørgensen, P. (2002), 'Challenging times for the Balkans', in Burdett (ed.),

Kabele, J. and Potuček, M. (1995), *The Formation and Implementation of Social Policy in the Czech Republic as a Political Process* (Vienna, Institute for Human Sciences).

Kaiser, M.J. (2000), 'Pareto-optimal electricity tariff rates in the Republic of Armenia', *Energy Economics* 22, 463–495.

Kazakevicius, E. et al. (1998), 'The residential space heating problem in Lithuania', *Energy Policy* 26, 859–872.

Kennedy, D. (1999), *Competition in the Power Sectors of Transition Economies*, EBRD Working Paper No. 41. (London, EBRD).

Kennedy, D. (2002a), *Liberalisation of the Russian Power Sector*, EBRD Working Paper No. 69. (London, EBRD).

Kennedy, D. (2002b), 'Regulatory reform and market development in power sectors of transition economies: the case of Kazakhstan', *Energy Policy* 30, 219–333.

Kirdar, U. and Silk, L. (eds) (1995), *People: From Impoverishment to Empowerment* (New York, New York University Press).

Kirsch, S. (1995), 'The incredible shrinking world? Technology and the production of space', *Environment and Planning D: Society and Space* 13, 529–555.

Klarer, J. (1997), 'Regional overview', in Klarer and Moldan (eds).

Klarer, J. and Moldan, B. (eds) (1997), *The Environmental Challenge for Central European Economies in Transition* (New York, Wiley).

Kočenda, E. and Cábelka, S. (1999), 'Liberalization in the energy sector in the CEE-countries: transition and growth', *Osteuropa-Wirtschaft* 44, 104–116.

Kopačka, L. (2000), 'Transformation of the Czech society and economy and energy industry', *Acta Universitatis Carolinae – Geographica* 29, 39–59.

Kornai, J. (1980), *Economics of Shortage* (Amsterdam, London, North-Holland).

Kostelecký, T. et al. (1997), *The Housing Market and Its Impact on Social Inequality: A Study of Prague and Brno*, SOCO Project Paper No. 54. (Vienna, SOCO (Social Consequences of Economic Transformation in East Central Europe), Institute for Human Sciences).

Kovačević, A. (2004), *Stuck in the Past: Energy, Environment and Poverty in Serbia and Montenegro* (Belgrade, United Nations Development Programme).

Kramer, J.M. (1991), 'Energy and the environment in Eastern Europe', in DeBardeleben (ed.),

Kreibig, U. et al. (2001), 'The power sector in Central and Eastern Europe: more competition needed in the run-up to EU membership', *Economic Bulletin* 38, 33–38.

Kreidl, M. (2000), 'Změny v percepci chudoby, bohatství a životního úspěchu [Changes in the perception of poverty, wealth and success]', in Matějů and Vlachová (eds).

Kuddo, A. (1995), 'How can poverty and inequality be avoided in Eastern Europe', in Kirdar and Silk (eds).

Kumssa, A. and Jones, J.F. (1999), 'The social consequences of reform in transitional economies', *International Journal of Social Economics* 26, 194–210.

Lackó, M. (1997), *Do Power Consumption Data Tell the Story? Electricity Intensity and the Hidden Economy in Post-Socialist Countries*, Working Paper. (Luxembourg, International Institute for Applied Systems Analysis).

Lampietti, J. and Meyer, A. (2002), *When Heat is a Luxury: Helping the Urban Poor of Europe and Central Asia Cope with the Cold* (Washington, D.C., World Bank).

Lavigne, M. (1999), *The Economics of Transition: From Socialist Economy to Market Economy* (Basingstoke, Macmillan).

Law, J. and Bijker, W. (1992), 'Postscript: technology, stability and social theory', in Bijker and Law (eds).

Lee, R. and Wills, J. (eds) (1997), *Geographies of Economies* (London, Arnold).

Legro, S. (1998), 'Making more by using less: energy efficiency in Russia', *BISNIS Bulletin* 7, 25.

Levine, M.D. et al. (1995), 'Energy efficiency policy and market failures', *Annual Review of Energy and the Environment* 20, 535–556.

Lewis, P. (1982), *Fuel Poverty Can Be Stopped* (Bradford, National Right to Fuel Campaign).

Leyshon, A. (1995), 'Missing words: whatever happened to the geography of poverty?', *Environment and Planning A* 27, 1021–1028.

Leyshon, A. and Thrift, N. (1995), 'Geographies of financial exclusion: financial abandonment in the United States and United Kingdom', *Transactions of the Institute of British Geographers* 20, 312–341.

Liggett, H. and Perry, D. (eds) (1995), *Spatial Practices* (London, Sage).

Lokshin, M. and Popkin, B.M. (1999), 'The emerging underclass in the Russian Federation: income dynamics, 1992–1996', *Economic Development and Cultural Change* 47, 803–829.

Lovei, L. et al. (2000), *Maintaining Utility Services for the Poor: Policies and Practices in Central and Eastern Europe and the Former Soviet Union* (Washington, D.C., The World Bank).

Lubinski, M. (1998), 'The Case of Poland: International financing and eco-funding', in Müller and Ott (eds).

Lux, M. (2000a), *The Housing Policy Changes and Housing Expenditures in the Czech Republic* (Prague, Institute of Sociology, Academy of Sciences of the Czech Republic).

Lux, M. (2000b), 'History and challenges of 'social' rental housing in the Czech Republic', in Lux (ed.),

Lux, M. (ed.), (2000b), *Social Housing in Europe* (Prague, Institute of Sociology, Academy of Sciences of the Czech Republic).

Lux, M. (2001), 'Social housing in the Czech Republic, Poland and Slovakia', *European Journal of Housing Policy* 1, 189–209.

Lux, M. et al. (2003), *Standardy Bydlení 2002/03: Finanční Dostupnost a Postoje Občanů [Housing Standards 2002/03: Financial Accessibility and Views of the Citizens]* (Prague, Academy of Sciences of the Czech Republic).

Mácha, M. and Woleková, H. (1998), 'Komparace vývoje státní sociální podpory v ČR a SR [Comparison of the development of state social support in the Czech and Slovak Republics', in Potůček and Rodičová (eds).

Machonin, P. (2000), 'Teorie modernizace a Česká zkušenost [Modernization theory and the Czech experience]', in Mlčoch et al. (eds).

Magnusson, L. and Ottoson, J. (eds) (1997), *Evolutionary Economics and Path Dependence* (Cheltenham, Edward Elgar).

Makkai, T. (1994), 'Social policy and gender in Eastern Europe', in Sainsbury (ed.),

Mandelbaum, M. (1978), *Flexibility in Decision Making. An Exploration and Unification* (Toronto, Department of Industrial Engineering, University of Toronto (Unpublished dissertation)).

Marangos, J. (2002), 'A political economy approach to the neoclassical model of transition', *American Journal of Economics and Sociology* 61, 259–276.

Martinot, E. (1998), 'Energy efficiency and renewable energy in Russia: transaction barriers, market intermediation, and capacity building', *Energy Policy* 26, 905–615.

Massey, D. (1984), 'Introduction: Geography Matters', in Massey and Allen (eds).

Massey, D. and Allen, J. (eds) (1984), *Geography Matters! A Reader* (Cambridge, Cambridge University Press).

Matějů, P. (2000), 'Subjektivní mobilita – Rekostrukce souvislostí mezi objektivním a subjektivním sociálním statusem [Subjective mobility – a reconstruction of the dependence between an objective and subjective social status]', in Matějů and Vlachová (eds).

Matějů, P. and Kreidl, M. (2000), 'Obnova vnitřní konzistence sociálně-ekonomického statusu [Renewing the internal consistency of socio-economic status]', in Matějů and Vlachová (eds).

McAuley, A. (1991), 'The economic transition in Eastern Europe: employment, income distribution, and the social security net', *Oxford Review of Economic Policy* 7, 93–105.

McCormick, J. and Philo, C. (1995), 'Where is poverty? The hidden geography of poverty in the United Kingdom', in Philo (ed.),

McDermott, G.A. (1997), 'Renegotiating the ties that bind: the limits of privatization in the Czech Republic', in Grabher and Stark (eds).

McDowell, L. (2004), 'Masculinity, identity and labour market change: some reflections on the implications of thinking relationally about difference and the politics of inclusion', *Geografiska Annaler B* 86, 45–56.

McMaster, I. (2001), *Privatisation in Central and Eastern Europe: What Made the Czech Republic so Distinctive?* (Glasgow, European Policies Research Centre, University of Strathclyde).

McMurrin, S. (ed.), (1980), *Tanner Lectures on Human Values* (Cambridge, Cambridge University Press).

Meessen, H. (1998), 'A practitioner's perspective', in Müller and Ott (eds).

MERM (Ministry of Economy of the Republic of Macedonia) (1998), *Pricing Methodology for Different Types of Energy* (Skopje, Government of the Republic of Macedonia).

Meyers, S. et al. (1994), 'Energy use in a transitional economy: the case of Poland', *Energy Policy* 22.

MFRM (Ministry of Finance of the Republic of Macedonia) (2002), 'Progress in Transition', *MF Bulletin* 2002, 12–35.

Milanović, B. (1991), 'Poverty in Eastern Europe in the years of crisis, 1978 to 1987: Poland, Hungary, and Yugoslavia', *World Bank Economic Review* 5, 187–205.

Milanović, B. (1999), 'Explaining the Increase in inequality during transition', *Economics of Transition* 7, 299–341.

Milbourne, P. (1997), 'Hidden from view: poverty and marginalisation in the British countryside', in Milbourne (ed.),

Milbourne, P. (ed.), (1997), *Revealing Rural 'Others': Representation, Power and Identity in the British Countryside* (London, Pinter).

Milková, M. and Vašečka, I. (1998), 'Srovnávací analýza systému sociální pomoci v České a Slovenské Republice po roce 1989 [Comparative Analysis of the Social Assistance Systems of the Czech and Slovak Republics After 1989]', in Potůček and Rodičová (eds).

Mingione, E. (1998), 'Fragmentation et exclusion: la question sociale dans la phase actuelle de transition des villes dans les sociétés industrielles avancées', *Sociologie et sociétés* 30, 1–15.

Mlčoch, L. (2000), 'Restrukturalizace vlastnických vztachů: institucionální pohled [The restructuring of property relations: an institutionalist view]', in Mlčoch et al. (eds).

Mlčoch, L. et al. (eds) (2000), *Economic and Social Changes in Czech Society After 1989: An Alternative View* (Prague, Charles University).

MLSA (Ministry for Labour and Social Affairs) (published online in 2005), 'Služby MPSV – Statistiky [Sevices of the MLSA – Statistics]', <http://portal.mpsv.cz/sz/stat>, accessed 20 October 2005.

MLSP (Ministry for Labour and Social Policy) (published online in 2005), 'Reforms of the Social Protection System of the Republic of Macedonia', <http://www.mtsp.gov.mk>, accessed 5 March, 2005.

Mohan, J. (2000), 'Geographies of welfare and social exclusion', *Progress in Human Geography* 24, 291–300.

Molnar, L. (2002), *Social and Environmental Impacts of Power Sector Reform in Hungary* (Roskilde, UNEP Risoe Centre on Energy, Climate and Sustainable Development).

MRD (Ministry for Regional Development) (2001), 'Housing Policy Concept', in

Muiswinkel, G.M. (1992), *Energy in Eastern Europe*, EIB Working Paper. (Luxembourg, European Investment Bank).

Müller, F. and Ott, S. (1998), *Bridging Divides – Transformation in Eastern Europe: Connecting Energy and Environment: Refocusing the Transformation Process.* (Baden-Baden, Nomos).

MUPC (Ministry for Urban Planning and Construction) (2000), *Prostoren Plan na Republika Makedonija – Nacrt [Draft Spatial Plan of the Republic of Macedonia]* (Skopje, Government of the Republic of Macedonia).

Musil, J. (1992), 'Recent changes in the housing system and policy in Czechoslovakia: an Institutional Approach', in Turner et al. (eds).

Nicita, A. and Pagano, U. (2001), *The Evolution of Economic Diversity* (London, Routledge).

Niggle, C. (1997), 'Changes in income distributions and poverty rates in Central European transitional economies', in Davidson and Kregel (eds).

North, D. (1990), *Institutions, Institutional Change and Economic Performance* (New York, Cambridge University Press).

Nove, A. (1992), *An Economic History of the USSR, 1917–1991* (Harmondsworth, Penguin).

Obadalová, M. (2000), 'K výdajům na bydlení v 90. letech [Housing expenditure during the 1990s]', *Sociální Politika* 12, 10–12.

Obadalová, M. and Vavrečková, V. (2000), *Náklady na bydlení českých domácností [Housing costs of Czech households]*, Working paper. (Prague, Research Institute for Labour and Social Affairs).

OECD (Organization for Economic Co-Operation and Development) (1994), *Energy Balances of Non-OECD Countries, 1991–1992* (Paris, OECD).

OECD (Organization for Economic Co-Operation and Development) (2003), *Energy Balances of Non-OECD Countries, 2000–2001* (Paris, OECD).

OECD/IEA (Organization for Economic Co-Operation and Development/ International Energy Agency) (1997), *Energy Efficiency initiative*, Volume 2: Country Profiles and Case Studies. (Paris, OECD/IEA).

Opitz, P. (2000), 'The (pseudo-) liberalisation of Russia's power sector: the hidden rationality of transformation', *Energy Policy* 28, 147–155.

Pallot, J. and Shaw, D. (1982), *Planning in the Soviet Union* (London, Croom Helm).

Paugam, S. (1995), 'The spiral of precariousness: a multidimensional approach to the process of social disqualification in France', in Room (ed.),

Pawley, M. (1997), *Terminal Architecture* (London, Reaktion Books).

Peck, J. (2000), 'Doing regulation', in Clark et al. (eds).

Peck, J. (2001), *Workfare States* (New York, Guiford Press).

Percy-Smith, J. (ed.), (2000), *Policy Responses to Social Exclusion: Towards Inclusion?* (Buckingham, Open University Press).

PEREEA (Protocol on Energy Efficiency Protocol and Related Environmental Aspects) (2001), *Regular Review – Draft* (Skopje, Government of the Republic of Macedonia).

Petroleum Economist (1997), *The Guide to World Energy Privatisation* (London, Petroleum Economist and Arthur Andersen).

PHARE Multi-Country Energy Programme (1995), *Study of Investment Climate in the Energy Sector in the Black Sea Region: Final Report* (Sofia, PHARE Multi-Country Energy Programme).

Philo, C. (ed.), (1995), *Off The Map: The Social Geography of Poverty in the UK* (London, Child Poverty Action Group).

Pichler-Milanovich, N. (1994), 'The role of housing policy in the transformation process of Central-East European Cities', *Urban Studies* 31, 1097–1115.

Pickles, J. and Smith, A. (eds) (1998), *Theorising Transition: The Political Economy of Post-Communist Transformations* (London, Routledge).

Pickvance, C. (1997), 'Decentralization and democracy in Eastern Europe: a sceptical approach', *Environment and Planning C* 15, 129–142.

PiEE (Power in East Europe) (1999), 'Hungary: new tariffs to reduce cross-subsidy', 1 April 1999

PiEE (Power in East Europe) (2002), 'Temelin plant enters operational phase', 2 September 2002.

Pierson, C. (1999), 'Continuity and discontinuity in the emergence of the welfare state', in Bryson et al. (eds).

Pisarev, A. (2006), 'Mafijata pali i žari [The mafia burns and chars]', *Forum plus*, 30 September 2006.

Plotnikova, M. (2002), *Regional Income Patterns in EU Accession Countries: an Exploratory Analysis*, Working Paper. (Urbana-Champaign, University of Illinois).

Polackova Brixi, H. et al. (1999), *Fiscal adjustment and contingent government liabilities: case studies of the Czech Republic and FYR Macedonia*, Policy Research Working Paper Series No 2177. (Washington, D.C., World Bank).

Pollert, A. (1988), 'Dismantling flexibility', *Capital and Class* 34, 42–75.

Pollert, A. (ed.), (1991), *Farewell to Flexibility?* (Macmillan, Basingstoke).

Polterovich, V.M. (1999), 'Institutsional'nye lovushki i ekonomicheskie reformy [Institutional traps and economic reforms]', *Ekonomika i Matematicheskie Metody* 35, 12–21.

Potter, R. and Unwin, T. (1995), 'Urban-rural interaction: physical form and political processes', *Third World Cities* 12, 67–73.

Potůček, M. (1999), *Not Only the Market: the Role of the Market, Government, and the Civic Sector in the Development of Postcommunist Societies* (Budapest, Central European University Press).

Potůček, M. and Rodičová, I. (1998), 'Porovnanie vývoja Českej a Slovenskej sociálnej politiky po roku 1989 [Comparison of the Developments of Czech and Slovak Social Policy After 1989]', in Potůček and Rodičová (eds).

Potůček, M. and Rodičová, I. (eds) (1998), *Sociální Politika v Čechách a na Slovensku po Roce 1989 [Social Policy in the Czech Republic and Slovakia After 1989]* (Prague, Karolinum).

Poulantzas, N. (1973), *Political Power and Social Classes* (London, Sheed and Ward).

Pridham, G. and Gallagher, T. (eds) (1999), *Experimenting With Democracy: Regime Change in the Balkans* (London, Routledge).

Průša, L. (2002), 'Analýza vývoje systému sociální pomoci v 90. letech [Analysis of the development of the social assistance system during the 1990s]', *Práce a Mzda* 2002, 29–37.

Przeworski, A. (1991), *Democracy and the Market: Political and Economic Reforms in Eastern Europe and Latin America* (Cambridge and New York, Cambridge University Press).

Raiser, M. et al. (2000), *The Measurement and Determinants of Institutional Change: Evidence From Transition Economies*, Working Paper No. 60. (London, EBRD).

Raiser, M. et al. (2001), *Social Capital in Transition: A First Look At the Evidence.*, EBRD Working Paper No. 61. (London, EBRD).

Ravallion, M. and Lokshin, M. (1999), *Subjective Economic Welfare*, World Bank Working Paper. (Washington, D.C., World Bank).

Redmond, G. and Hutton, S. (2000), 'Poverty in transition economies: an introduction to the issues', in Hutton (ed.),

REGNUM News Agency (2006), 'According to Odessa expert, 'Ukraine is at the verge of a second energy crisis'', 14 February, 2006.

Reijnders, J. (1997), *Economics and Evolution* (Cheltenham, Edward Elgar).

RIA Novosti (2006), 'Russian finance minister upbeat about economic development, WTO entry', 16 July 2006.

Rimashevskaya, N. and Voitenkova, G. (2000), 'Impoverishment and social exclusion in Russia', in Hutton (ed.),

Rodionov, P.I. (1999), *Toplivno-Energeticheskiy Kopmpleks Rossii: Ekonomicheskoe Regulirovanie [The Fuel and Energy Complex of Russia: Economic Regulation]* (Novosibirsk, Nauka).

Room, G. (ed.), (1995), *Beyond the Threshold: The Measurement and Analysis of Social Exclusion* (Bristol, The Policy Press).

Room, G. et al. (1989), 'New poverty in the European Community', *Policy and Politics* 17, 165–176.

Rudge, J. and Fergus, N. (eds) (1999), *Cutting the Cost of Cold: Affordable Warmth for Healthier Homes* (London, E and FN Spon).

Russia and CIS Business and Financial Daily (2006), 'Oil, gas, base for world energy in 21st century – G8 energy ministers', 16 March 2006.

Rutkowski, J. (2001), 'Earnings mobility during the transition: the case of Hungary', *Moct-Most* 11, 69–89.

Sachs, J. (1990), 'Eastern Europe's economies: what is to be done?' *The Economist*, 3 January 1990.

Sainsbury, D. (ed.), (1994), *Gendering Welfare States* (London, Sage).

Saith, R. (2001), *Capabilities: The Concept and Its Operationalisation*, QEH Working Paper Series No. 66. (Oxford, Queen Elizabeth House).

Sassen, S. (1991), *The Global City* (Princeton, Princeton University Press).

Schlack, R.F. (1996), 'Economics of transition: hypotheses toward a reasonable economics', *Journal of Economic Issues* 30, 617–627.

Schmieding, H. (1993), 'From plan to market', *Weltwirtschaftliches Archiv* 128, 216–316.

Schöpflin, G. (1991), 'Post-Communism: constructing new democracies in Eastern Europe', *International Affairs* 67, 235–251.

Secrest, T.J. (2002), *A Framework for Financing Small-Scale Energy Efficiency Projects in the Transition Countries of Central and Eastern Europe* (Richland, PNNL (Pacific Northwest National Laboratory)).

Sen, A.K. (1980), 'Equality of what?' in McMurrin (ed.),

Sennett, R. (1998), *The Corrosion of Character: The Personal Consequences of Work in the New Capitalism* (London, Norton).

SEVEn (The Energy Efficiency Center) (1999), *Analýza Dotací v Energetice [Analysis of Energy Subsidies]* (Prague, SEVEn).

SEVEn (The Energy Efficiency Center) (2001), *Bulletin Energetýckich a Ekonomýckich Informací za Českou Republiku [Energy and Economic Bulletin for the Czech Republic]* (Prague, SEVEn).

Sibley, D. (1998), 'Social exclusion and the Roma in transition', in Pickles and Smith (eds).

Sidaway, J.D. and Pryke, M. (2000), 'The strange geographies of emerging markets', *Transactions of the Institute of British Geographers* 25, 187–202.

Sik, E. and Redmond, G. (2000), 'Coping strategies in Central European Countries', in Hutton (ed.),

Siner, M. and Stern, J. (2001), 'Reform of the electricity market in transition economies: how to avoid traps of deregulation', *World Bank Transition Newsletter* 11, 8–10.

Smith, A. (1997), 'Breaking the old and constructing the new? Geographies of uneven development in Central and Eastern Europe', in Lee and Wills (eds).

Smith, A. (2002), 'Culture/economy and spaces of economic practice: positioning households in post-communism', *Transactions of the Institute of British Geographers* 27, 232–250.

Smith, A. and Pickles, J. (1998), 'Introduction: theorising transition and the political economy of transformation', in Pickles and Smith (eds).

Smith, A. and Swain, A. (1998), 'Regulating and institutionalising capitalisms: the micro-foundations of transformation in East-Central Europe', in Pickles and Smith (eds).

Smith, D.M. (1992), 'Geography and social justice: some reflections on social change in Eastern Europe', *Geography Research Forum* 12, 128–142.

Sobolewski, M. and Zylicz, T. (2000), 'Reforming environmental and energy policies in the economic transition process', in Jasinski and Pfaffenberger (eds).

Sojka, M. (2000), 'Ten years of Czech transformation: transformation, inequality, and integration', in Mlčoch et al. (eds).

Solomon, P. and Foglesong, T.S. (2000), 'The two faces of crime in post-Soviet Ukraine', *East European Constitutional Review* 9, 72–76.

Speak, S. and Graham, S. (2000), 'Service not included: marginalised neighbourhoods, private service disinvestment, and compound social exclusion', *Environment and Planning A* 31, 1985 – 2001.

SSO (State Statistical Office) (1993), *Statistical Yearbook of the Republic of Macedonia* (Skopje, SSO).

SSO (State Statistical Office) (1995), *Census of Population, Households, Dwellings and Agricultural Holdings in the Republic of Macedonia in 1994* (Skopje, SSO).

SSO (State Statistical Office) (1996), *Household Expenditure in the Republic of Macedonia in 1995* (Skopje, SSO).

SSO (State Statistical Office) (2004), *Household Expenditure in the Republic of Macedonia in 2003* (Skopje, SSO).

SSO (State Statistical Office) (2005), *Household Expenditure in the Republic of Macedonia in 2004* (Skopje, SSO).

Stark, D. (1992), 'The great transformation? Social change in Eastern Europe', *Contemporary Sociology* 21, 299–304.

Stenning, A. (1997), 'Economic restructuring and local change in the Russian Federation', in Bradshaw (ed.),

Stenning, A. (2003), 'Shaping the economic landscapes of post-socialism? Labour, workplace and community in Nowa Huta, Poland', *Antipode* 35, 761–780.

Stenning, A. (2005), 'Post-socialism and the changing geographies of the everyday in Poland', *Transactions of the Institute of British Geographers* 30, 113–127.

Stern, J. (1994), 'Economic regulation in Central and Eastern Europe', *Economics of Transition* 2, 391–398.

Stern, J. and Davis, J.R. (1998), 'Economic reform of the electricity industries of Central and Eastern Europe', *Economics of Transition* 6, 427–460.

Stiglitz, J.E. (1994), *Whither Socialism?* (Cambridge, Mass., and London, MIT Press).

Stojmilov, A. (1995), *Geografija na Republika Makedonija [Geography of the Republic of Macedonia]* (Skopje, Prosvetno Delo).

Strickland, C. and Sturm, R. (1998), 'Energy efficiency in World Bank power sector policy and lending: new opportunities', *Energy Policy* 26, 873–883.

Struyk, R.J. (1996), 'Housing privatization in the former Soviet bloc to 1995', in Andrusz et al. (eds).

Summers, L. (1992), *Keynote Address: Knowledge for Effective Action*, Proceedings of the World Bank Annual Conference on Development Economics. (Washington, D.C., World Bank).

Sunega, P. (2002), *Adresný Příspěvek na Nájemné v Prostředí České Republiky: Komparace Vybraných Modelů [Rent Subsidies in the Czech Republic: A Comparison of Selected Models]* (Prague, Institute of Sociology, Academy of Sciences of the Czech Republic).

Sykes, R. et al. (eds) (2001), *Globalization and European Welfare States: Challenges and Change* (New York, Macmillan).

Sýkora, L. (1996), 'The Czech Republic', in Balchin (ed.),

Sýkora, L. (2003), 'Between the state and the market: local government and housing in the Czech Republic', in Lux (ed.),

Sýkora, L. et al. (2000), 'Changes in the spatial structure of Prague and Brno in the 1990s', *Acta Universitatis Carolinae Geographica* 35, 61–76.

Szajkowski, B. (1999), 'Macedonia: an unlikely road to democracy', in Pridham and Gallagher (eds).

Szelényi, I. (1983), *Urban Inequalities Under State Socialism* (Oxford, Oxford University Press).

Szomolányová, J. et al. (1999), *Energetická Strategie České Republiky [Energy Strategy of the Czech Republic]* (Prague, SEVEn (The Energy Efficiency Center)).

Telgarsky, J.P. and Struyk, R.J. (1990), *Toward a Market-Oriented Housing Sector in Eastern Europe* (Washington, D.C., Urban Institute Press).

Tembo, F. (2003), *Participation, Negotiation and Poverty: Encountering the Power of Images* (Aldershot, Ashgate).

Timofeev, A.V. et al. (1998), 'Characteristics of the financial position of electric utilities', *Studies of Russian Economic Development* 9, 544–547.

Todorovska, V. (2007), 'Albania experiences most severe electricity crisis in history', *A1 news*, 31 January 2007.

Toplifikacija (2005), *The District Heating System of the City of Skopje* (Skopje: Toplifikacija.

Torrey, B.B. et al. (1999), 'Income transitions in Central European households', *Economic Development and Cultural Change* 47, 237–257.

Townsend, P. (1979), *Poverty in the United Kingdom: A Survey of Household Resources and Standards of Living* (London, Allen Lane).

Trajkovski, I. (2002), *National Human Development Report for Macedonia 2002* (Skopje, United Nations Development Programme).

Turner, B. et al. (eds) (1992), *The Reform of Housing in Eastern Europe and the Former Soviet Union* (London, Routledge).

UNDP (United Nations Development Programme) (1998), *Poverty in Transition* (Sofia, UNDP).

UNDP (United Nations Development Programme) (2003), *Avoiding the Dependency Trap* (Bucharest, UNDP).

UNDP (United Nations Development Programme) (2005), *Human Development Report 2005. International Cooperation at a Crossroads: Aid, Trade and Security in an Unequal World* (New York, UNDP).

UNECE (United Nations Economic Commission for Europe) (1991), *Energy Reforms in Central and Eastern Europe* (Brussels, UNECE).

UNGA (United Nations General Assembly) (1985), *The International Bill of Human Rights* (New York, United Nations).

UNICEF (United Nations Childrens' Fund) (2000), *After the Fall: The Human Impact of Ten Years of Transition* (Florence, UNICEF).

Ürge-Vorsatz, D. et al. (2006), 'Energy in transition: From the iron curtain to the European Union', *Energy Policy* 27, 2279–2297.

USAID (United States Agency for International Development) (2002), *Saving Energy in the Czech Republic* (Washington, D.C., USAID).

USDoE (United States Department of Energy) (2002), *An Energy Overview of the Czech Republic* (Washington, D.C., USDoE).

Utrinski Vesnik (2000), 'Najskapo so struja, najevtino so parno [Electricty most expensive, district heating cheapest fuel]', 9 October 2000.

Večerník, J. (1995), *Household Incomes and Social Policies: The Czech Republic From 1989 to 1995*, Project Paper No. 25. (Vienna, Institute for Human Sciences).

Velody, M. et al. (2003), *A Regional Review of Social Safety Net Approaches in Support of Energy Sector Reform* (Washington, D.C., US Agency for International Development).

Voigt, S. and Engerer, H. (2000), *Institutions and Transformation – Possible Policy Implications of the New Institutional Economics*, Working Paper: New Developments in Economic Research. (Berlin, German Federal Ministry of Finance).

von Hirschhausen, C. and Opitz, P. (2000), *Power Utility Re-Regulation in East European and CIS Transformation Countries: An Institutional Interpretation.* (Dundee, Centre for Energy, Petroleum, Mineral Law, and Policy).

von Hirschhausen, C. and Wälde, T.W. (2001), 'The end of transition: an institutional interpretation of energy sector reform in Eastern Europe and the CIS', *Moct-Most: Economic Policy in Transitional Economies* 11, 93–110.

Vorsatz, D. (1997), 'Lessons from the Hungarian energy sector sell-out', *Transition: the World Bank Newsletter About Transition Economies* 8, 12.

Vromen, J.J. (1995), *Economic Evolution. An Enquiry Into the Foundations of New Institutional Economics* (London and New York, Routledge).

Wacquant, L. (1999), 'Urban marginality in the coming millennium', *Urban Studies* 36, 1639–1647.

Wälde, T.W. and von Hirschhausen, C. (1998), *Regulatory Reform in the Energy Industry of Post-Soviet Countries*, CPMLP Paper No. 3/98. (Dundee, Centre for Petroleum and Mineral Law and Policy).

Wicks, M. and Hutton, S. (1986), *Electricity Use by Families With Children* (London, Electricity Consumers' Council).

WIIW (Vienna Institute for International Economic Studies) (2005), *South-East Europe – A Region in Competition for FDI* (Vienna, WIIW).

Wills, J. (2004), 'The third way, welfare, work, and local governance reform', *Environment and Planning A* 36, 571–578.

World Bank (1993), *The World Bank's Role in the Electric Power Sector: A World Bank Policy Paper* (Washington, D.C, World Bank).

World Bank (1998), *Profile of Energy Sector Activities of the World Bank in Europe and Central Asia Region* (Washington, D.C., World Bank).

World Bank (1998), *FYROM – Joint Country Assistance Strategy* (Washington, D.C., World Bank).

World Bank (1999a), *Non-Payment in the Electricity Sector in Eastern Europe and the former Soviet Union*, World Bank Technical Paper No. 423. (Washington, D.C., World Bank).

World Bank (1999b), *Privatization of the Power and Natural Gas Industries in Hungary and Kazakhstan*, World Bank Technical Paper No. 451. (Washington, D.C., World Bank).

World Bank (1999c), *Former Yugoslav Republic of Macedonia: Focusing on the Poor*, Report 19411-MK. (Washington, D.C., World Bank).

World Bank (2000), *The Road to Stability and Prosperity in South Eastern Europe: a Regional Strategy Paper* (Washington, D.C., World Bank).

World Bank (2005), 'News and Broadcast: Left Out in the Cold', 5 May 2005.

Wynn, M. and Wynn, N. (1979), *Prevention of Handicap and the Health of Women* (London, Routledge and Kegan Paul).

Yavlinsky, G. and Braguinsky, S. (1994), 'The inefficiency of laissez-faire in Russia: hysteresis effects and the need for policy-led transformation', *Journal of Comparative Economics* 19, 88–116.

Yeung, H. (2005), 'Rethinking relational economic geography', *Transactions of the Institute of British Geographers* 30, 37–51.

Zaostrovtsev, A. (2000), *Rent Extraction in a Rent-Seeking Society*, SPIDER Working Paper No.10. (St. Petersburg, St.Petersburg State University).

Zoon, I. (2001), *On the Margins: Roma and Public Services in Romania, Bulgaria, and Macedonia. With a Supplement on Housing in the Czech Republic* (New York, Open Society Institute).

Index